ACPL ITEM DISCARDED

OCEANS AND CONTINENTS IN MOTION

OTHER BOOKS BY H. ARTHUR KLEIN

Surf-Riding

The New Gravitation: Key to Incredible Energies

Holography: With an Introduction to the Optics of Diffraction, Interference, and Phase Differences

Fuel Cells: An Introduction to Electrochemistry

Bioluminescence: A First Look at Light from Life

Masers and Lasers

Surfing

Graphic Worlds of Peter Bruegel the Elder

Hypocritical Helena
—Translated and edited from the works of Wilhelm Busch

Max and Moritz
—Translated and edited from the works of Wilhelm Busch

BY H. ARTHUR KLEIN AND MINA C. KLEIN

Käthe Kollwitz: Life in Art

Israel: Land of the Jews—A Survey of Forty-three Centuries

Temple Beyond Time: The Story of the Site of Solomon's Temple

Peter Bruegel the Elder, Artist of Abundance

Great Structures of the World

Surf's Up! An Anthology of Surfing

INTRODUCING MODERN SCIENCE

OCEANS AND CONTINENTS IN MOTION

AN INTRODUCTION TO CONTINENTAL DRIFT AND GLOBAL TECTONICS

by H. Arthur Klein

J. B. LIPPINCOTT COMPANY /
PHILADELPHIA AND NEW YORK

U.S. Library of Congress Cataloging in Publication Data

Klein, H   Arthur.
Oceans and continents in motion.

(Introducing modern science)
SUMMARY: Traces the development of the continental drift theory and examines insights about the earth's history that have resulted in a new science called global tectonics.
1. Continental drift—Juvenile literature. 2. Geology, Structural—Juvenile literature. [1. Continental drift. 2. Geology, Structural] I. Title.
QE511.5.K56   551.4'1   72-3731
ISBN-0-397-31271-7

Copyright © 1972 by H. Arthur Klein
Member, Authors League of America
   and National Association of Science Writers, Inc.
All rights reserved
Printed in the United States of America
First Edition

To the memory of the poet Robinson Jeffers (1887–1962) who comprehended so many of the events that shaped the little planet he found "very beautiful."

1723888

I gazing at the boundaries of granite and spray ... felt behind me
Mountain and plain, the immense breadth of the continent, before
   me the mass and doubled stretch of water.

                                                      ... there is in me
... the eye that watched before there was an ocean.

That watched you fill your beds out of the condensation of thin
   vapor and watched you change them,
That saw you ... wear your boundaries down, eat rock, shift
   places with the continents.

                              —from "Continent's End" in *Tamar and Other Poems* (1924)

Contents

	Acknowledgments	9
	Author's Note	11
1	Curiously Conformable Continents	15
2	Drift Ideas Move Onward	24
3	Alfred Wegener at Work	35
4	Dry Land and Ocean Bottoms—Deep-Seated Differences	43
5	Continental Cross Sections and Sub-Crustal Cookery	53
6	The Mountain-Making Mother . . .	68
7	Bumping Blocks and Quaking Crusts	77
8	Convections, Conductions, and Temperature Differences	92
9	Radioactive Insights—and an Era of Eclipse	102
10	Magnetic Magic	111
11	Creating, Carrying, and Consuming Crusts Under the Oceans	120
12	Total Tectonics	134
13	Patterns and Processes	148
14	Applying the New Insights	172
	Index	183

Acknowledgments

> Bliss was it in that dawn to be alive,
> But to be young was very heaven!

So wrote the poet William Wordsworth early in the nineteenth century as he recalled his feelings during days he had spent in France of the Revolution, 1791–92.

Something of the same high excitement radiates from the current revolution of insights in the earth sciences. It is a revolution wrought largely by young, or at least young-minded, geophysicists. Their ideas seem to make great leaps, literally earth-embracing in import, as the following book seeks to show.

The writer recalls now, however, that this book—his sixth for the Introducing Modern Science series—did not seem to leap toward completion. Sometimes progress seemed almost as imperceptible as the continental drift it describes. Yet the process would have been still less rapid, but for the help of a number of scientists. Their comments, corrections, or encouragements merit the gratitude here expressed.

They are, however, quite clear of blame for such errors of fact or interpretation as the book may contain. The author alone bears that brunt. Sincere thanks are offered to the following scientists based outside the boundaries of the U.S.A., beginning with the most distant: Dr. H. Takeuchi of the University of Tokyo, a conversation with whom was most helpful in many ways; Drs. Xavier Le Pichon, Centre Océanologique de Bretagne, Brest, France; Stanley K. Runcorn, Professor of Geophysics, University of Newcastle upon Tyne, England; and J. Tuzo Wilson, Principal, Erindale College, Toronto, Canada.

Likewise, thanks go to a number of earth scientists based in the United States, listed here alphabetically: Drs. William E. Benson, Head, Earth

Sciences Section, National Science Foundation, Washington, D.C.; Kenneth S. Deffeyes, Geology Department, Princeton University; David T. Griggs, Professor of Geophysics, University of California at Los Angeles; Leon Knopoff, Professor of Geophysics, University of California at Los Angeles; W. Jason Morgan, Geology Department, Princeton University; William A. Nierendorf, Professor of Physics and Chancellor of Marine Sciences, Scripps Institution of Oceanography, La Jolla, California; Donald L. Turcotte, Professor of Engineering, Cornell University, N.Y.; and Roland E. Von Heune, Geologist, U.S. Geological Survey, Menlo Park, California.

For permissions to use pictures, diagrams, maps or quotations, thanks go also to: M. R. Cracraft, Jr., Amoco Production Company, Tulsa, Oklahoma; Harriet Ewing, Lamont-Doherty Geological Observatory of Columbia University, Palisades, N.Y.; Nelson Fuller, Scripps Institution of Oceanography; Vicki Groninger, Associate Editor, *Engineering Quarterly*, Cornell University; Dr. Herman L. Hoeh, Ambassador College, Pasadena, California.

Also to Mary E. Luken, Geological Society of America, Boulder, Colorado; R. L. Rahder, editor, *SpaN*, magazine of the Standard Oil Company of Indiana; Ralph Segman, NOAA Environmental Research Laboratories, Denver, Colorado; and A. F. Spilhaus, Jr., American Geophysical Union, Washington, D.C.

Dover Publications, Inc. generously permitted use of both illustrations and quotations from their 1966 edition of *The Origins of Continents and Oceans* by Alfred Wegener. Donnan C. Jeffers of Carmel, California permitted the use of several relevant quotations from the poems of Robinson Jeffers. Lines from "Continent's End" in *Tamar and Other Poems* appear by permission of Random House.

Repeatedly, the fine facilities and helpful atmosphere of the Geology Library at UCLA were used, for which warm thanks are offered to John D. Hill, librarian of that collection, and his staff.

Finally, as so often before, the author thanks Mina C. Klein for services too manifold to specify but unfadingly gratifying to recall.

Malibu, California, 1972

Author's Note

Measurements and quantitative estimates are stated here mainly in the standard and accepted units for all scientific work—the metric and SI (International System) units. However, we in the United States are still saddled with the obsolescent and obscure "common units." Hence, equivalents in inches (in), feet (ft), or miles (mi) are often inserted in parentheses following the metric amounts.

Though it is a formality rather than a fundamental, the author has tried to observe recommended practice regarding abbreviations and capitalization of the accepted units. Thus we have the meter (m), centimeter (cm), kilometer (km), the gram (gm), the kilogram (km); also the watt (W), kilowatt (kw), and megawatt (MW). And for temperature we have both the degree Celsius (°C) and the Kelvin (K), which is an exact equivalent, when temperature differences are being stated.

The recommended use of the singular, rather than plural, for quantities greater than one unit is also observed. Thus we write 500 kilometer (not kilometers), 33 kilogram (not kilograms), and so forth.

Heat energy is expressed in the unit called the joule (J), rather than the once-common calorie. The latter is, of course, the gram-calorie, not the weight-watcher's calorie (the kilogram-calorie)! Heat *power,* the rate of flow of heat energy, is expressed in the watt (W).

For those who prefer to compare the new with the old-time units, here are approximate equivalents: 1 kw equals 57 BTU (British

Thermal Units) per minute; 1 MW equals 57,000 BTU per minute, or 950 BTU per second. Also, 1 kw equals 44,250 foot-pound per minute, or 738 foot-pound per second. Finally, 1 joule equals 0.24 gram-calorie; and 1 BTU equals 1055 joule.

OCEANS AND CONTINENTS IN MOTION

1. Curiously Conformable Continents

This is a book about mysteries, cryptic clues, and daring detective work. The detectives, however, are not tracking down criminals. Rather, they are tracing the past whereabouts of entire continents.

Their quarry is, in fact, our planet earth, home to all humankind and our fellow forms of life. Our story of global detection begins with a situation much like one familiar to readers of the many fictions about international spies and secret operatives. That situation runs about like this:

A new agent, to be known only as A, is assigned to a perilous mission, M, in a far-off nameless nation N. The first step must be to make contact with the local operator there, an unknown called just U. Then together A and U will do the dark deeds that they have to do.

"But how shall I find and recognize U?" asks A. In answer, his chief, S, the sophisticated head of the Secret Intelligence Bureau, tears off a jagged corner from the cover of a book of safety matches, and tosses the remainder to A, saying, "In the capital city of N, you will enter about noon the side door of the Rundown Café just across from the City Hall. If someone approaches you, asking for a light, you will offer this matchbook, and say softly, 'Sorry, this seems a trifle the worse for wear . . .'

"If that someone then asks, 'Would this possibly be what's missing?' and hands you the missing corner—and if it fits beyond doubt, then that someone is your contact U!"

"But," says *A* with the skepticism common to new agents, "what if that piece does *not* fit . . . ?"

In answer *S* only nods darkly but unmistakably toward the bulge beneath *A*'s armpit . . .

In our tale of tectonics and wandering continents, the torn-apart portions are as enormous as the matchbook is tiny. They are indeed the continents and subcontinents, no less. But the same rules apply. If great coastal curves and lesser indents and bumps on widely separated masses of land fit better than could have been caused by mere chance—then one conclusion becomes clear: at some past time they must have been rent asunder and moved to their present positions.

Comparisons of widely separated contours can be made, however, only when the shapes of the land masses have been well measured and mapped. Such graphic information about the earth's major features is rather recent in human history. After all, only five or six thousand years have passed since "prehistoric" times, even though the earth itself attained just about its present bulk and size some four to five *billion* years ago.

In fact, barely five centuries have passed since humans became able to draw and analyze the coastlines and other great features of continents clearly enough to make the torn-matchbook kind of comparisons.

One world finds another—Just before the end of the fifteenth century, Europeans "discovered" the New World, including North and South America, the connecting neck of land called Central America, and various fertile and desirable nearby islands. Among the Europeans who arrived were some with skill in measuring and making miniatures (maps) of the surface of the earth. By about the mid–sixteenth century, fairly complete sets of such maps were engraved and published in northern Europe. Outstanding among these "atlases" were those of two great Flemish leaders in the new geography and cartography: Gerhard Mercator (1512–1594) and Abraham Ortelius (1527–1596). Their maps pictured large sections of the Atlantic coastlines of both North and South America as well as of the continents of Europe and Africa.

In Britain during the early years of the seventeenth century such maps were studied with keen attention, especially by one ambitious, far-sighted, and original thinker: Francis Bacon (1561–1626). In them he found evidence to support the great ideas which he urged upon his contemporaries. Bacon was, in fact, the prophet of the method of *science* that was yet to be born in practice.

He had a radical program: If men would only study "nature" persistently, precisely, even quantitatively, then they could gain such mastery over nature that life could be made safer, more abundant, better worth living. He was clear about the road to be followed: "Nor can nature be commanded except by being obeyed, and so those twin objects, human knowledge and human power, do really meet in one; and it is from ignorance of causes that operation fails. . . ."

Conformable instances—In 1620 Bacon's masterpiece was published. It was titled *Novum Organum,* which can be translated as "new method" or even "new program." Writing in Latin, language of scholars and philosophers of that time, Bacon called attention to some examples of the kind of nature study he had in mind.

Most important for the present book, he pointed for the first time in print to curious likenesses between the shapes of huge land masses widely separated on earth. Specifically, Bacon compared the Atlantic coastlines of the continents of Africa, "stretching to the Straits of Magellan," and of South America, which to him was "the region of Peru."

Each of these opposite coastlines, he found, had "similar isthmuses and similar promontories, which can hardly be by accident."

Bacon called this a "conformable instance"—a case of significant conformity or matching of shapes. This was the very sort of thing, in fact, that his new method proposed should be studied with keenest care, for the useful knowledge that could be extracted from it.

Both Africa and South America, this philosophical iconoclast insisted, were "broad and extended toward the north," but "narrow and pointed toward the south." They showed, indeed, a multiple kind of conformity.

But Bacon was more a method-maker than a scientist in any modern sense of that word. He drew no specific conclusions and suggested no hypotheses to account for these conformable shapes. However, during the three and a half centuries since Bacon, notable thinkers and scientists have grappled with these and similar "conformable instances." Out of those efforts have arisen truly revolutionary insights into the sciences of the earth. Those insights are sometimes called the *continental drift* theory, but more and more often since the late 1960s they are grouped under another intriguing name—*global tectonics,* or *plate tectonics.* An introduction to these breathtaking insights is the aim of this book.

Placet on continental placements—Within some thirty years after Bacon died, a curious work appeared in France. Its author, R. P. François Placet, told his readers that before the great global flood recorded in the Bible in the book of Genesis, America was not separated from "the other parts of the world." Both North and South America, Placet proposed, had been continuous with or contiguous to Europe and Africa.

The teasing conformity of shape that Bacon had noted called for the fitting of the easternmost part of Brazil into Africa's Gulf of Guinea. Those matching parts are now separated by some 4,600 kilometers (2,900 miles). Hence, a vast movement must have taken place in post-flood times, movement involving what Bacon had called the "greater parts" of the globe.

After François Placet, nearly a century and a half passed before a competent student of the earth's physical structure returned to this bold idea. Alexander von Humboldt (1769–1859), a German geographer, explorer, naturalist, and educator, pioneered in fields today called *physical geography* and *geophysics*. Between 1709 and 1804 he journeyed widely in South and North America and studied the enormous current in the southern Pacific Ocean, which is still called the Humboldt Current in his honor.

In 1800 Humboldt suggested that the Atlantic Ocean had been born as a kind of gigantic riverbed, and that flowing waters had carved out the coastlines of the continents that border it.

"Our Atlantic Ocean," Humboldt wrote, "exhibits every sign of a valley structure. It seems as if the rush of the waters had directed its thrust first toward the northeast, then northwest, and then once more northeast. Supporting this bold opinion are the parallelisms of the coasts north of 10° South [latitude], the projected and retracted parts, the convexity of Brazil opposite the Gulf of Guinea, the convexity of Africa below the same latitude as the Gulf of Antilles."

In fact, on opposite sides of this Atlantic valley, Humboldt noted that "retracted coastlines, rich in islands, face convex coasts on the [opposite] bank."

A Frenchman, Antonio Snider-Pellegrini, took another step with his book, whose title in English means *The Creation and Its Mysteries Unveiled*. Appearing in the late 1850s, the book contained the first hypothetical drawing showing how the continents once may have clustered together. South and North America nested against Africa and Europe respectively, while the island continent of Australia fitted into a niche along the east African coastline, near today's Tanzania and Kenya.

Such bold reshuffling of continental sites was based mainly on discoveries of identical or similar fossils in widely scattered beds of coal on both sides of the Atlantic. Coal is the compressed and carbonized remains of vegetation that ages ago grew lush in tropical climates. Yet today's coal beds are found where the climate may be quite cool or even absolutely arctic.

A survey of shapes—In 1857 a learned periodical in Edinburgh, Scotland, published a study by W. L. Green, with an extended title, "The Causes of the Pyramidal Form of the Outlines of the Southern Extremities of the Great Continents and Peninsulas of the Globe." Green wrote in a markedly modern manner about "segments of the earth's crust which float on [the earth's] liquid core."

Then in 1875 appeared Green's book, *Vestiges of the Molten Globe*. These "vestiges," he declared, were the modern continents, floating on earth's core, which—he believed—was still liquid and molten.

In that same year the Russian scientist Y. Bykhanov again called attention to the striking conformity of the shapes of the eastern (African) and western (South American) coasts of the Atlantic.

Several other writers came forward with theories that entire continents shift position more or less continually in relation to the portions of the earth that lie under them. Some of these theorists supposed that the continents did not move much relative to one another, but that as a whole they shifted over the internal layers of the earth.

In 1880 at Zurich, Switzerland, appeared a German work by H. Wettstein, also with a lengthy title: in English it means *The Currents of the Solid, Liquid, and Gaseous Forms and Their Meaning for Geology, Astronomy, Climatology, and Meteorology*. Wettstein pictured the continents being altered in shape as they shifted in relation to the layers of earth under them.

Our earth spins ever eastward. Hence Wettstein argued that the gravitational attraction of the sun operates to pull the continents *westward,* dragging them across the viscous substances that support them. Wettstein thus saw tidal forces as the motive power for great continental motions.

He supposed, however, that the present ocean beds were really sunken continents. He was unaware of the fact that existing continents do not suddenly end where they meet the ocean, and that their continental shelves slope far out under the oceans.

During the 1870s the astronomer-physicist G. H. Darwin (eldest son of the great biologist Charles Darwin) worked, together with Osmond Fisher, on a dramatic hypothesis, which could be called "the earth-birth of the moon." They proposed that the moon, whose mass is less than 1.3 percent of that of the earth, was originally thrown off from the swifter-spinning earth that existed eons ago. They suggested also that the centrifugal force supplied by the earth had been assisted by the gravitational attraction of a passing star—an attraction strong enough to raise up a great wave of hot earth-substance. When the top of that wave broke away, it became the moon, which gradually moved outward to its present distance, some 380,000 km (236,000 mi) from its parent earth.

The great depression left by the loss of this moon-matter became, Darwin and Fisher said, the bed of the Pacific, giant among oceans. During the early 1880s, Fisher amplified this vivid idea. He suggested that after the moon had flown off, the remaining continents of earth must have been shifted into new positions.

The American poet Robinson Jeffers (1887–1962), familiar with many ideas of modern science, wrote a poem, "The Great Wound," published in 1964. The wound Jeffers referred to was the Pacific Ocean, beside which he lived and wrote. He imagined, as he stood on his cliff, the sound of "the half-molten basalt and granite tearing apart" and the spectacle of the moon "leaping up" toward that passing star.

Today the idea of the earth-birth of the moon has been given up. In fact, analysis of the rocks brought back to earth from the moon by astronauts has shown clearly that the two bodies must have had different origins. Our satellite could not have been an offspring or a spin-off from our planet. But well into the twentieth century, the earth-birth theory of the moon's origin may have had more supporters among geologists than did the continental drift theory.

In 1907, in fact, William H. Pickering (1858–1938), of a family that has included several prominent American scientists, published a paper, "Place of Origin of the Moon—the Volcanic Problem." He linked the Darwin-Fisher hypothesis to the striking intercontinental fit of the coasts on both sides of the South Atlantic. Pickering suggested that the same cataclysm that tore the moon from the earth also pulled the Americas away from Europe-Africa. The New World, he suggested, moved from the Old to help fill the void the moon had left. Thus was the bed of the present Pacific formed. And as a further consequence, the gap newly created between Europe-Africa on one side and the Americas on the other also filled with seawater. It became the present Atlantic Ocean.

Continental drift theories today agree that the opening of the Atlantic began *not more* than 200 to 250 million years ago. The Pickering-Fisher hypothesis of the formation of the Atlantic in a still-molten earth had to be at least 2,500 million years too early in its timing.

In fact, one of the surprises in continental drift theory is the relative recentness of the major movements that gave the earth its present facial features. When one becomes used to geological time scales of billions of years, a mere 150 million years begins to seem like "only yesterday."

However, ideas about the earth's true age have altered even more than ideas about its shape and position in space. Only a few generations ago, most people, if they thought at all about the age of their earth, were likely to turn to confident estimates by Bible scholars. For example, there was Bishop Ussher, who precisely dated the Creation at 4004 B.C., or just about 5,976 years before these lines were published!

The Jewish calendar is based on a chronology only a little different from that offered by Ussher. This book is being written in the year 5,732 of the Jewish calendar. The age of the earth, as seen by modern geology and geophysics, must be very nearly one million times as great as these theological estimates. Yet it is not the sheer age of earth that seems most impressive. The *how* and the *why* prove even more wonderful than the *when,* as we look deeper into the events that almost certainly first formed and then shaped and reshaped our earth.

2. Drift Ideas Move Onward

Early in this century, two Americans gave important support to the idea that great continents were once united, but moved apart.

In 1908, F. B. Taylor stressed significant similarities in major mountain systems. The *Bulletin of the Geological Society of America* in 1910 published his "Origin of the Earth's Plan," emphasizing what Taylor termed the "tertiary" mountain belt—meaning mountains formed some 70 million years ago.

Another specialist in the geology of mountain-making, Howard B. Baker, found clues in many parts of the world. These led him during nearly a score of years, starting in 1911, to propose that in the relatively recent geologic past all the continents had been nested together. Only thus could he fit together closely corresponding mountain groups and reconstruct the continuous mountain chain which, he believed, must have existed when these mountains were first formed.

Today, more than sixty years later, geologists and geophysicists trace great mountain chains through the present continents and the island systems in offshore oceans (Figures 2-1 and 2-2). Some of the most important of these, however, must have been formed later than those on which Taylor and Baker concentrated.

Taylor's writings have been praised as persuasive and "elegant." The Tertiary period which he stressed began about 70 million years ago and continued until a mere million years ago. During this period much folding of layers or strata near the earth's surface took place,

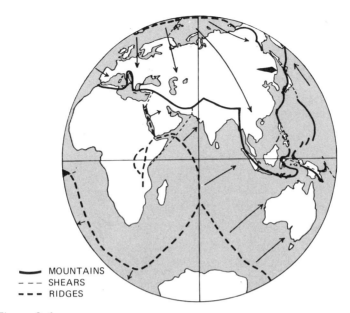

──── MOUNTAINS
– – – SHEARS
▬ ▬ ▬ RIDGES

Figure 2-1.

Relatively young mountain chains curve around the entire earth. They are shown here and in Figure 2-2 by heavy solid lines.

The heavy dashed lines show the global system of ridges or rises. From these, new crustal matter emerges and spreads slowly in both directions, perpendicular to the general line of the ridge.

Light dashed lines suggest shears, or faults, along which the earth's great tectonic plates slip past each other.

Arrows show the general direction of the present major plate movements. Where two sets of arrows encounter or oppose each other, either head-on or at an angle, mountains are likely to be found. This suggests how mountain-making, drifting of continents, and spreading of ocean floors are all interrelated.

Here the mountain line begins at the Atlas Mountains of northern Africa's Atlantic coast. The mountains run eastward, cross the Mediterranean to Sicily, then run up the narrow peninsula of Italy into the Alps of northern Italy and Switzerland.

A sharp southward turn leads down to the Greek peninsula and archipelago. Then the mountain line moves east and south through the Zagros Mountains of Iraq and Iran, to the huge Himalayas in the north of India. Here the tectonic plate carrying the subcontinent of India has pushed below that which carries Asia-Europe.

Next, a sharp southward turn takes the mountain line through the Naga Hills of Burma, the Andean Islands, Sumatra, Java, and—after a "hairpin turn"—northward through Celebes.

Then the mountain line "hops," island by island, east of the mainland of Asia: the Philippines, Taiwan, the islands of Japan, and the Kuril Islands.

Next, in an easterly direction, the mountain curve crosses the Bering Sea.

Indicated also on this view is a separate and smaller mountain line through New Guinea, north of Australia; then through the islands of Palau, Yap, Guam, the Marianas, Iwo Jima, and Bonin. This sector terminates in a short curved shear line to the major Japanese island of Honshu.

Whereas these mountain lines appear as ranges and peaks on the continents, they take the form of separated islands, or island chains, when they run through the ocean.

Reproduced by permission of J. Tuzo Wilson, University of Toronto.

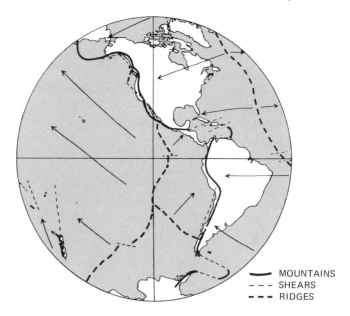

Figure 2-2. From Asia the globe-girdling mountain chain crosses the northern Pacific in the form of the Aleutian Islands. Then it swings along southern Alaska and down the Pacific coast through Canada and the states of Washington, Oregon, and California, all the way to the tip of Lower (Baja) California.

Here it crosses over the widening Gulf of California and reappears on the mainland of Mexico and Guatemala, to the Caribbean Sea at the Gulf of Honduras. A gap again appears at the Caribbean, but the line is resumed in the short arc of islands of the West Indies. Next, the chain seems to leap back to the mainland of South America, at Point Gallinas, west of the Gulf of Venezuela.

Thence it runs, enormous and unbroken, along the mighty Cordillera de los Andes, through Peru and Chile, down to Cape Horn at the tip of the continent.

From there a "shear line" runs east and south. It leads to a resumption of the heavy mountain line at South Georgia Island and the South Sandwich Islands. Another shear line leads from there back to the northern tip of the curving Palmer Peninsula, and thence toward the major mass of the Antarctic continent. Not until about 20° from the South Pole does that global mountain line terminate.

This imposing sequence of relatively youthful mountains and mountains-become-islands does not include all mountains on earth. There are others. But it does include the great majority of the mountains that seem to have been formed by recent opposing or clashing motions of the earth's great tectonic plates.

Reproduced by permission of J. Tuzo Wilson, University of Toronto.

resulting in mountain systems that have not yet eroded or weathered away. Only *after* these mountains were folded, Taylor believed, did the continents of today pull apart from one another. Thus he pictured Greenland splitting from North America while the land masses of the Americas moved westward, ever more distant from Europe-Africa, and while part of that land mass sank under the waters that filled the new and widening Atlantic Ocean.

Enter Alfred Wegener—Just about the time Taylor and Baker made their contributions, a young German geophysicist and meteorologist began grappling with the same provocative problems. He was Alfred Wegener (1880–1930), born in Berlin and educated at universities in Germany and Austria. (Figure 2-3).

Wegener was no boy prodigy. He had worked hard and long in his chosen fields. In 1906, aged twenty-six, he had taken part in a Danish expedition into northeast Greenland. It went on for two years, amid many rigors and dangers. Wegener returned, reported on meteorological findings in Greenland, and became a lecturer in meteorology and astronomy at the well-known University of Marburg.

Figure 2-3. Alfred Wegener (1880–1930), meteorologist, arctic explorer, and a pioneer proponent of continental drift theory in modern times.
Photograph by Deutsche Fotothek Dresden.

In 1910, already thirty years old, he was first struck—much as Bacon had been—by the shape similarities of the "coastlines on either side of the Atlantic." This led him naturally to consider the notion of a possible continental drift. However, by his own report, Wegener at first gave no serious attention to that idea—"because I regarded it as impossible."

In 1911, however, he happened to read a scientific work telling of marked similarities between older forms of animal life in West Africa, on the one hand, and South America, on the other. That report, in the effort to account for these close likenesses, suggested that long ago a "land bridge" must have linked the two continents.

Wegener knew enough earth science to find such a supposition even more improbable than continental drift itself. Today earth scientists generally agree that the land bridge idea is untenable and that present ocean floors were never, and could never have been, dry land that later sank below sea level.

However, how could continental drift be reconciled with the growing body of facts gathered by many separate sciences dealing with the earth and its life forms? Wegener became fascinated—even, perhaps, obsessed. During the last months of 1911 he worked with a kind of calm fury. By the end of that year he had accepted the continental drift idea, and from then on he defended and advanced it with energy and ingenuity.

He could have found reason enough to keep his drift conclusions pretty much to himself. He had worked closely with an eminent German authority on meteorology, Professor Wladimir Köppen of Hamburg, who urged Wegener to make himself a well-grounded specialist in meteorology and not scatter his interests too widely. But Wegener was "somewhat obstinate in great matters," as a friend later said of him. Surely, the origin of the continents and the oceans was a great matter, if ever there was one.

Before the first week of 1912 had passed, Wegener had entered the public arenas on behalf of the continental drift theory. In the famous old German City of Frankfurt am Main he lectured on "The Development of the Large-Scale Features of the Earth's Crust, [Considered] on a Geophysical Basis."

A few days later, in Marburg, he talked before a scientific society on "Horizontal Displacement of the Continents." This was a formal but fitting name for the drift theory. Academic and orthodox geologists had long taught that *vertical* displacements of the earth's crusts constantly occurred, as in the formation of mountains, buttes, raised plains, etc. But the notion of long-range "horizontal" cruises by entire continents—that was something else again! And this mere university lecturer who proposed it was not even a professor, not even a proper geologist. Just a meteorologist, whose bag should have been the winds and the weather!

Modestly enough, but without apology or pause, Wegener went on minding what was not his meteorological or astronomical business. Before the end of 1912, his first published paper on continental drift appeared in *Petermanns Mitteilungen,* an established scientific journal. Its title was as simple as some of the others in the past had been elaborate: "The Origin of the Continents." In that same active year, Wegener also published another paper on the same general subject in the *Geologische Rundschau.*

Wegener's first two papers on drift totaled only forty pages, but with them he laid the foundations on which he later built his book and a renown that has survived a temporary eclipse.

Ardent drift defenders—In lectures and in conversations with his students at Marburg University, Wegener unfolded the expanding evidence that he assembled in support of the controversial continental drift concept. In many cases these students became active supporters of his views. One of his friends later recalled that they "would not have hesitated to employ their fists as an argument" in the effort to convince someone who opposed the drift idea.

Meanwhile, Wegener by no means abandoned his own specialty of meteorology. Before 1912 ended, he was off on another of the difficult scientific expeditions in the Arctic. In the frigid interior of Greenland he became one of the first Europeans to survive a winter on a high Arctic glacier. After that ordeal he traveled across the Greenland ice cap, a journey of some 1,200 km (750 mi).

Following his return to Germany, Wegener, then in his early thirties, married the daughter of his teacher and mentor, Professor Köppen, the eminent meteorologist of Hamburg.

The outbreak of World War I in the summer of 1914 forced Alfred Wegener reluctantly into uniform as a lieutenant in a field regiment of the Imperial German Army. He "took his military service very hard," according to a friend. "Not because of the dangers and hardships—which rather appealed to a nature such as his—but because of the difficult conflict it must have caused, between his duty to his Fatherland and his innermost conviction: that war was futile."

According to this friend, Wegener was one of the rare people who regard life's true purpose as the advancement of the well-being of mankind as a whole, not just of one people or nation.

As the German army advanced through Belgium on its way to France, Wegener was wounded by a shot in the arm. After he had recovered enough to be returned to active duty, he was again wounded—this time in the neck.

These injuries were not complete misfortunes for Wegener. For one thing, they took him off the list of those classed as *KV*—the German term meaning fit for combat service. Also, during periods of enforced convalescence Wegener found time, in spite of the tensions and dangers of war, to return to his great work on continental drift. During 1915 he used "a prolonged sick leave" to expand and perfect his previous material.

Origin of Wegener's "Origin"—The result was the first (1915) edition of the book whose English title became *The Origin of Continents and Oceans*. This initial German edition was only ninety-four pages long.

Wegener finally was given behind-the-lines tasks in the German army's meteorological (weather) service. When at last the war ended with the defeat of Germany late in 1918, he returned to his interrupted triple career in research, teaching, and exploration. By 1920 he had prepared a second edition of his *Origins* book, nearly half again as long as the first.

Wegener continued to update and enlarge that work in a third edition in 1922 and a fourth and final one in 1929. In these revisions he tried to review and reply to objections raised by many opponents of his views. He took pains to include the results of recent researches in varied fields of science relevant to drift theory.

During a dozen years after the end of World War I, Wegener was recognized more and more as the principal protagonist and expounder of drift theory. His influence reached beyond the limits of the German-speaking regions. In 1924, French, Spanish, and English versions of his book were published. The following year a Russian-language edition appeared in the Soviet Union.

Work on bygone climates—By this time Wegener had co-authored another book. Collaborating with his father-in-law, Köppen, he brought out a survey of the complex subject of *paleoclimatology,* the study of climates that prevailed in past geological periods. It was, in Wegener's words, "a collection of geological and paleontological material."

His continental drift studies were thus not the only ones in which he delved deep into the remaining traces of conditions in the distant past. In summarizing the conclusions of this new book, Wegener noted that continental drift was only "one of many" causes found for the climate changes over long periods of time in the great land masses of the earth. In fact, regarding the most recent periods, Wegener readily conceded that continental drift might not even have been among the foremost causes of major climatic changes. Nevertheless, all that Wegener and Köppen learned about ancient climates did harmonize with drift theory in general.

During the years after the end of World War I, Wegener was employed as a department chief in the marine observatory (at Hamburg) of the new German Republic. At the same time he lectured at the University of Hamburg, but for many years he did not reach the rank of professor, so sought after in German scientific and academic circles.

He became a professor at last in 1924 when he was in his midforties. A chair in the twin fields of meteorology and geophysics was created especially for him at the University of Graz, Austria. The linking of these two fields reflected his own style of work. Wegener tried to escape the overspecialization and pigeonholing that then afflicted the sciences of the earth.

Finale of a fighter—Even after he became a *Herr Professor,* Wegener remained devoted to on-site exploration and observation. He planned still another expedition to carry on his earlier work in Greenland. Prepared in 1929, the expedition set out in 1930. It achieved some important results, including measurements showing that the depth of the ice crust over the interior of Greenland exceeds 1.8 km (1.1 mi).

There, on October 30, 1930, just fifty years old, Wegener set out on a dog-sled journey to his glacier camp near the west coast. No one ever saw him again. Almost certainly his remains lie somewhere in the ice he had measured and, till then, known how to master. Alfred Wegener's untarnishable memorial remains, beyond doubt, his firm, patient, and resourceful advocacy and advancement of the continental drift theory, and his book on that subject remains a key to the early evidence for that theory and to Wegener as thinker.

It is not so brilliantly written as some other classics of science history. In many ways it is more a patient compilation than a startlingly original revelation. The years since 1929 have dated much of it, for in the meantime vast advances have been made in many areas. Nevertheless, the book still packs into surprisingly little space much evidence and interpretation to support drift theory. And to a considerable extent it aids the understanding of advances made long after Wegener wrote it.

In the foreword to his final edition, Wegener issued a challenge to his colleagues and contemporaries: "Scientists still do not seem sufficiently to understand that *all* of the earth sciences must contribute evidence revealing the state of our planet in earlier times." Elsewhere he again emphasized that *all,* in declaring that the truth about earlier states of the earth could be attained only "by combining all this evidence."

The following pages summarize some of this *all* and show the extent to which Wegener himself realized his own demand: that every earth science be combined in the great effort to reconstruct periods long past.

3. Alfred Wegener at Work

Wegener stressed that his book, *The Origin of Continents and Oceans,* was intended equally for "geodesists, geophysicists, geologists, paleontologists, zoogeographers, phytogeographers, and paleoclimatologists."

For specialists in each of these many technical subdivisions of the earth sciences, Wegener set out to summarize the evidence for continental drift that originated in specialties other than their own. He aimed quite consciously to serve as a sort of scientific go-between or mediator, crossing the boundaries that separated the specialties. He went to work, in other words, in a way that today is called "interdisciplinary" and is much praised as a method for escaping overspecialization.

Beyond doubt, his way of working made many of his scientific colleagues and contemporaries uneasy or even resentful. They were unwilling to have their settled ideas upset by evidence brought in from outside the boundaries of the specialties in which they held sway.

Wegener dealt with an impressive breadth of material.

Geodesy applies mathematics, and especially geometry, to determinations of the size and shape of the earth and its major parts, including the dry land and ocean areas.

Geophysics studies the physical characteristics of the earth and its component materials, both crustal (surface) and interior. It includes measurements or estimates of temperature, heat content, pressure, elasticity, viscosity, plasticity, tensile strength, density, and so on.

Geology was the third specialty listed by Wegener. Geology is, or seeks to be, the study of the physical history of the earth, the substances (especially rocks) that compose it, and the changes that have taken place or are still taking place on and in it. Inevitably, geology is involved with enormous time intervals. It should be, according to Wegener, closely integrated with geophysics. The lack of such integration he regarded as a weakness of the geology prevalent around him in the 1920s.

The prefix *paleo-* always means "from long ago." *Paleontology* studies life forms that existed in bygone geologic periods. *Paleoclimatology* studies the climates that prevailed then. Since we have neither embalmed animals nor weather bureau records from those periods millions of years ago, the evidence must come from the remains—fossils and other traces—of living things which then existed, or from rocks and sediments that have endured from then to now.

Climate is broader and more general than *weather.* Climatologists study and compare prevailing and average weather conditions. They deal with such variables as temperature, humidity, atmospheric pressure, precipitation (rain and snow), sunshine or cloudiness, and even with the composition, alteration, or pollution of the atmosphere. Hence climatology in recent times has concerned itself much with the effects of smog and other pollutants belched into the atmosphere of earth by myriads of automobiles, trucks, and industrial establishments.

Animal, vegetable, and mineral—Wegener worked, quite deliberately, with sciences dealing with all three major kingdoms or categories of the "things" on earth—mineral, vegetable, and animal. For instance, *zoogeography* studies the distribution and relationships of animals on earth now or in the rather recent past. *Paleontology,* however, studies, from surviving traces, the living creatures of periods long past. *Phytogeography* studies the distribution of plants in the present and recent past.

In spite of this scope and breadth, Wegener's historic book remained rather compact. Even its latest and largest edition had fewer than 250 pages. Using many diagrams and references to papers published in various science journals, Wegener summarized what he believed to be most important and enlightening. He reminded his readers again and again that further research and study were needed and would undoubtedly alter many of his tentative interpretations.

The main theme or moral of the book is stated quite simply: *The continents must have shifted.* Wegener reconstructed the routes they seemed to have followed in this shifting. Thus South America and Africa, once side by side and unified, split in two. Wegener was specific about the shape-matching which had caught the keen eyes of Francis Bacon. "The large rectangular bend formed by the Brazilian coast at Cap São Roque," he noted, exactly matches "the bend in the African coast at the Cameroons." Furthermore, south of those corresponding contours, "every projection on the Brazilian side precisely matches a congruent bay on the African side," and so on.

Once South America and Africa had split, they drifted further apart, "like pieces of a cracked ice floe on water." Thus wrote the experienced Arctic explorer, thoroughly familiar with the way ice floes floated, strained, split, severed, and drifted asunder. Even today Wegener's ice floe analogy is still used to explain more graphically how the great crustal plates of earth move and interact.

According to Wegener's estimates, the great continental splitting and separating began about 30 million years ago. The present extent of the separation is about 6,000 km (3,730 mi). Hence, Wegener estimated the average speed of the drift apart had been about 20 centimeters (7.9 inches) per year.

Today we have a different and more dependable way of dating

rocks, based on radioactive measurements. The new time estimates are about 180 to 200 million years, instead of Wegener's 30 million. His average speeds of drift thus are reduced to about 3 cm (1.2 in) per year.

Other enormous tear-aparts—Severing and separating of Africa–South America was but one of the vast drift events described by Wegener. North America, he declared, once lay alongside and formed part of Europe—or rather of Eurasia, for Europe geographically is but a kind of northwesterly projection of the enormous land mass of Asia.

Included in the Eurasian–North American block was also present-day Greenland, "at least from Newfoundland and Ireland northwards." This other great block, Wegener estimated, broke up a mere 5 to 7 million years ago—but not all at once. He visualized the breakup process as multiple, for as recently as 1 million years ago the westbound mass of North America split once again, leaving behind the sub-block that became the Greenland that Wegener had learned to know so well.

Wegener's time estimates provided only 2 to 4 million years for producing the separation of about 3,000 km (1,860 mi) between Newfoundland and Ireland. The new radioactive dating techniques require that those estimates be raised to between 40 and 80 million years, and the average rate of separation becomes about 3.5 to 8 cm per year. Today's best estimates of the average rates of such continental driftings are summarized simply though roughly as *the length of a man within a man's lifetime.* To a man or woman that must seem slow. To a geologist or geophysicist it seems very speedy indeed.

A subcontinent in a supercontinent—The supercontinent of Africa–South America, as reconstructed by Wegener, included also the present continents of Antarctica, Australia, and even the so-called subcontinent of India, now attached to the mainland of Asia. The name of *Gondwanaland* was given to this hypothetical continental collection. The name came from a past of India whose geology includes remarkable formations linking its past with that of parts of Africa to the south. It is often shortened to just Gondwana.

The giant Gondwanaland continent was not all dry land, however, for Wegener described much of it as being covered by shallow waters —mere seas or giant lakes rather than true oceans.

Wegener dated the splitting of Gondwanaland into sub-blocks during the Triassic, Cretaceous, and Jurassic geological periods. The resulting parts, in his picturesque words, then "drifted away in all directions." (This recalls a description in a sketch by Stephen Leacock, an economics professor who became a professional humorist, of the knight who hastily mounted his horse "and rode madly off in all directions!")

Modern geological dating places the start of the Triassic period at about 230 million years ago, the Jurassic period at about 180 million, and the Cretaceous period at about 130 million, with a duration of about 60 million years—that is, until some 70 million years ago.

Alterations en route—Wegener offered three maps to show how the world may have looked at three key points in the long process of continental splitting and drifting.

Surprising and spectacular changes in vertical as well as horizontal directions took place during these long drifts across the face of the earth. For instance, Wegener indicated that at one time the subcontinent of India had been connected to southern Asia by a "long stretch of land," so low and flat that it was covered by shallow waters— again, a sea rather than a true ocean.

However, India continued drifting northward toward Asia. As it moved, "this long junction zone became increasingly folded." The result—earth's mightiest mountain system, the Himalayas, plus "the many folded [mountain] chains of upland Asia."

In the 1970s, the specialists in continental drift, now widely called "global tectonics," describe the events that lifted the Himalayas in terms somewhat different from those used by Wegener. But they, too, agree that these greatest mountains on earth were formed because the vast crustal (tectonic) plate on which India rode north drove into, and then passed below, the greater plate which carried Asia. Mighty mountain systems indeed are caused by continental collisions and compressions, much as automobile crashes cause dents, folds, and ripples in the "rigid" metal skins, or bodies, of the cars involved.

According to Wegener's book, other massive mountain systems also were uplifted by such relatively recent continental movements. Thus he pictured the westward movements of both North and South America as strongly compressing their (western) leading edges. As Wegener visualized it, these continents, being pushed through the resisting floor of the Pacific Ocean, developed great folded mountain systems. Hence the enormous mountain ranges running from south to north for thousands of kilometers, such as the mighty Andes of South America and the mountain chains of western North America.

Wegener regarded the ancient floor of the Pacific Ocean as being especially resistant to the motions of the continental masses, for that floor was cold—"deeply chilled, hence a source of viscous drag." It was thus particularly unyielding, on the principle woven into the old folk saying, "Slow as molasses in January!"

More modern insights have deeply modified this part of Wegener's concept of drift processes. Wegener regarded the Pacific Ocean floor as far older than the continents which came drifting westward across it. At least it was older in the sense of having been there far longer than the continental intruders. However, we shall see that there is good reason to regard the ocean floors as younger, by far, than the continents that rise above them. Also, in important instances the ocean floor and the continent that it meets are moving *together* as parts of the same giant tectonic plate or raft.

Nevertheless, the newer theories also recognize mountain-making as a result of the same general processes that produce the great continental drifts. Wegener was correct so far as this connection was concerned, even if some details of the "machinery" may be different from those he visualized.

Island origins—The formation of great island chains and arcs in various parts of the world, especially in the western Pacific, was attributed by Wegener to the same basic drift processes. Thus he concluded that the Australian continental block, including what is now New Zealand and New Guinea, had split off long ago from what is now Antarctica, the continent around the South Pole.

The way Wegener saw it, New Zealand was at first in front of what became the east coast of Australia. The resistance of that cold old Pacific Ocean floor produced foldings, hence mountains, within New Zealand. Later, however, the direction of New Zealand's drift changed, cutting off those up-front mountains and leaving them lagging behind in the form of island chains, visible on every world map.

Wegener saw those islands simply as the summits of mountains emerging above the surface level of the Pacific. Similarly, he told how mountain chains on the easternmost margin of the Asian mainland had split off from the mainland and become island sequences whose curves clearly duplicated those of the mainland west of them.

Wegener wrote also that what is now Central America, linking North and South American continents, was once a land mass lying far to the east of its present position. Then, as it was moved westward through the old, resistant bed of the Pacific, it left behind it the island groups now called the Greater and Lesser Antilles. Even now, one may trace on a globe of the world the general direction, from northwest to southeast, in which slender Central America lies—and see that it is roughly parallel to a line running from Cuba to Haiti–Dominican Republic to Puerto Rico.

4. Dry Land and Ocean Bottoms — Deep-Seated Differences

In his book's first chapter, entitled "The Nature of Drift Theory," Alfred Wegener firmly emphasized the fundamental unlikeness between the continents on the one hand and the true ocean floors on the other. Drift theory itself, he insisted, proceeds "from the supposition that the sea-floors and the continents consist of *different* materials."

By *ocean floors* he meant the stone crusts or layers beneath such sediments as may have accumulated under the ocean waters. The ocean floor rock was really, in his view, "the free surface of the next layer inward." By inward he meant *downward,* toward the center of the earth. The ocean floor, then, was a layer that lay below, or downward from, the continental crusts, where such crusts existed.

Some very basic questions arise here. Why indeed does the face of our earth reveal both oceans and continents? If the earth originated as a hot, molten ball, the lighter substances could be expected to float to the top, or outside, while the heaviest would sink to the bottom, or center.

The waters of the oceans are the lightest—least dense—of the major layers now at earth's surface under the atmosphere. Next in density are the granitelike rocks typical of the continental crusts. Then, denser still, are the basaltlike rocks of the ocean floor crusts. And densest of all these near-surface substances are the so-called ultrabasic rocks of the earth's *mantle,* which lies deeper than both the continental and the ocean floor crusts.

A great paradox resulted from all this, and until recently it was never properly solved by geology. Why do not all these layers of differing densities lie symmetrically and concentrically one over the other, with the least dense (water) outside, and the most dense (ultrabasic mantle) within? The result would be an onionlike earth—mantle rock inside, then a shell or layer of the basalt rocks typical of ocean floors, followed by a shell or layer of continental-type granite, topped perhaps by a sedimentary layer, and finally on the outside, the waters of the universal ocean. Such a symmetrical earth would reveal no continents, no dry land showing, nothing but the waves of that universal ocean, whose uniform depth would be about 2.6 km (1.6 mi) everywhere.

That is, of course, the kind of earth we do *not* have. But *why* not? Only in great folk-tales of a worldwide flood, such as that which Noah and his household survived in the Ark, do we find even the supposition that the globe was ever totally covered by water.

Land-living creatures, including humankind, are made possible only by the puzzling fact that there are distinct and stable formations of dry land rising up out of the oceans. The key to this puzzle has been provided only by the tectonic concepts which developed out of continental drift theories. We owe insights as basic as this to Wegener and those who followed in his wake.

A separated earth—The separation of the dry land from the oceans is so striking that its origins are included in most of the great creation stories. Thus the biblical book of Genesis tells that, on the third day of creation, the Lord decreed this basic separation: "Let the waters under the sky be gathered into a single basin so that the dry land may appear." (Those are the words of a fine, clear and recent translation, *The New American Bible.)*

The very first ocean of the earth may indeed have been an enormous single, undivided one, covering more than seven-tenths of the earth's surface. Today the earth has more than one ocean basin, however. The familiar phrase about the "seven seas" is excessive, but oceanographers and physical geographers recognize at least three principal oceans at present, each fairly well defined by changes in crustal materials at its margins and marked by striking geological events taking place along its bottom.

The Pacific Ocean is by far the largest single feature of our earth, for it covers about 33 percent of the surface of the planet. Less than half as extensive is the Atlantic, with some 16 percent of the earth's surface. Only a little smaller is the Indian Ocean, with about 14 percent.

Maps are often marked to show an "Arctic Ocean." Actually, it should be called a *sea,* though it covers about 3 percent of the earth's surface. An additional 5 percent is occupied by other seas and watery sites, not true oceans.

Together, the oceans and the seas cover some 71 percent of the entire surface of the planet. Though so enormous in coverage, this "hydrosphere" is really little more than a thin film of water over the ocean floors. The volume of all the waters of the oceans and seas represents only about one-eighth of 1 percent of the entire bulk of the planet! The average density of the earth as a whole is about five and one-half times the density of pure water. The mass (or "weight") of the ocean waters amounts to only about one-fiftieth of 1 percent of the total mass of the earth as a whole.

Yet this comparatively tiny amount of water is spread out so thinly and broadly that the earth has very nearly two and one-half times as much watery surface as land surface!

Dry land in detail—Contrasting with the three true oceans are six continents, each one of which has less total area than the Indian Ocean, smallest of the oceans.

Largest of the continental masses is Eurasia. Including Asia, Europe, and the islands that clearly belong to them, it covers about 10 percent of the earth's total area. Next is Africa, with about 5.8 percent. Then come North America and its associated islands, with about 5.8 percent; followed by South America with about 3.5 percent. Antarctica, draped around the South Pole, has about 2.7 percent of earth's area. Last and least is Australia, the "island continent," with only 1.5 percent of the total area of earth.

If, as Wegener declared, the oceans occupy basins whose materials and construction differ basically from the slabs of continental crust, those differences should have much to tell us. They do indeed. They shed light on the earth's great past, its present, and its probable future.

The facts about the ocean floors, as marshaled by Wegener, run quite counter to a "natural" or "common sense" supposition: that the only real difference between the ocean floor and the dry land consists in the fact that the latter happens to lie high enough to project above sea level, while the former does not. Were this true, then it would only be necessary to drain away enough ocean water and the earth would acquire vast new areas of normal land, much as the Dutch, by draining their famous one-time Zuyder Zee, reclaimed large areas of precious land for farming and human habitation, leaving only the much reduced IJsselmeer.

Wegener showed, however, that the topography or contours of the true ocean bottoms proves that they had a different origin and character than the continents.

How deep?—In Shakespeare's *Romeo and Juliet,* Mercutio, fatally hurt by a sword, jests bitterly about his wound: " 'Tis not so deep as a well, nor so wide as a church door; but 'tis enough, 'twill serve." We have noted the width or extent of the major land and water areas of the world, but have yet to examine how deep, or high, both ocean floors and dry land elevations tend to be.

Sea level is the usual standard by which both are measured—the land altitudes as distances *above,* the ocean depths as distances *below* average sea level. The extremes of such upward and downward measurements may seem large to us humans, who seldom grow to heights greater than 2 meters. However, on the global scale the actual differences from sea level remain rather tiny.

Thus, there is less than 20 km (12.5 mi) difference between the highest height of the Himalayas and the lowest depth of the Mariana Trench at the bottom of the Pacific. This extreme difference is less than one-third of 1 percent of the radius of the earth itself.

The earth is a slightly oblate sphere, not perfectly smooth, but by no means rough or wrinkled, comparatively speaking. It is just about as irregular as a ball 1 m (39.3 in) in diameter whose highest surface point stands less than 1.7 millimeters above its lowest.

Yet even within these rather minor differences, earth's altitudes are so distributed that they lead to a surprising result. Wegener stated it this way: "In all geophysics there probably is hardly another law so clear and reliable as this one—that there are two preferential levels for the world's surface which . . . occur side by side and are represented respectively by the continents and by the ocean floors."

The statistics of land heights and subocean depths revealed to Wegener that this pattern of varied elevations could not have come from upliftings and sinkings of just "one original level." There had to be two separate origins. This is illustrated (Figure 4-1) by a so-called hypsometric curve showing the proportions of the earth's total surface that lie at various elevations above and depths below sea level. The two pronounced spikes or peaks of this curve tell an unmistakable story. The one at the right (L) shows that a strikingly large percentage of land lies at or quite close to a very low elevation (about 100 m above sea level). The peak at left (O) shows that a very large part of all ocean bottom lies a good way below sea level (at or close to 4.7 km below).

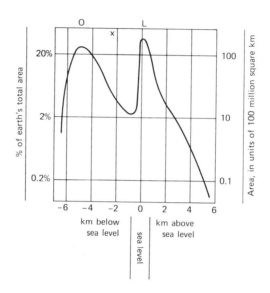

Figure 4–1. The tell-tale two-peaked curve, revealing that continents and ocean floors cannot have had a common origin. This "hypsometric" graph shows approximate proportions of the total area of earth (510 million km^2) at various depths below (left) and elevations above (right) sea level (center). The peak marked O shows the ocean floors' "preference" for the depth of about 4.7 km below sea level. Peak L shows the marked "preference" of land for the small elevation of about 100 m above sea level.

Based on data from E. Kossinna, *Die Erdoberfläche* (The Surface of the Earth), 1933.

Had both kinds of surface evolved from a common general level by shifting upward or downward, the curve would look quite different. It would be bell-shaped, with a single peak at about the point

Dry Land and Ocean Bottoms—Deep-Seated Differences 49

marked X on the graph, for 2.5 km below sea level is the approximate arithmetic average (or mean) level for *all* the earth's solid surfaces, totaling 510 million square km.

About 29.1 percent of earth's surface is above sea level, and of this "land" part, nearly seven-tenths is elevated less than 1 km. The other 70.9 percent of the planet is under water, and of this great part, more than three-quarters lies more than 3 km (but less than 6 km) below sea level.

Less than 5 percent of the earth's total area lies between 2 and 3 km below sea level, even though that range includes the average elevation of all solid surfaces of the earth.

Using his understanding of the separate origins and characteristics of continents and ocean floors, Wegener presented a simplified diagram to show a typical continental plate floating in the denser rock of the ocean crust, much as the ice floes that he often watched in Greenland float in the denser water in which they form. (Figure 4-2).

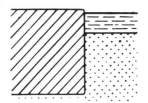

Figure 4–2. Continental crust (diagonal shading), being less dense, floats in the ocean floor crust (dotted shading) in this symbolic diagram by Alfred Wegener. The water of the ocean itself is indicated at upper right by horizontal dashes.

This diagram does not indicate the continental shelves, which taper out for some distance under the waters of the ocean before the bottom makes the final plunge down to the true oceanic depths.

Reproduced by permission of Dover Publications, Inc.

Any floating object sinks to a depth which depends on the ratio between the object's density and the greater density of the substance in which it floats. Arctic ice, being about 88 percent as dense as the seawater around it, floats about 88 percent below and 12 percent above that water. However, a block of wood only 60 percent as dense as the water would float 60 percent below and 40 percent above the water level.

These principles apply also to the great blocks of continental rock floating in the still denser rock of the ocean floor material, which, according to Wegener, consists of "denser, heavier material than that of the continental blocks." Such density differences are rooted in the different chemical compositions and crystalline structures of the two kinds of rock—continental and suboceanic.

Continental blocks are composed mostly of granitelike rocks in which silicon and aluminum compounds are prominent. The first syllables of the names of those two elements were put together to form a new name, *sial,* coined as a label for all such granitelike rocks.

On the other hand, the denser ocean floors are formed from basaltic rocks in which *si*licon and *ma*gnesium are prominent; hence the name *sima,* coined for all rocks in this general category. The sial continents float in the sima ocean floor material.

Because continents had moved so far, Wegener assumed that the sima, though dense rock stuff, over long periods of time must behave in sufficiently fluid fashion to allow the continental floes or blocks to be pulled slowly through it. The basaltic sima is dense, solid, and rigid enough, when tested by brief forces. But it must yield when subjected to sufficiently long-lasting forces, like those exerted during entire geological periods.

Denser still—Wegener was aware that important rock structures might be even denser than the sima. He cited geophysical studies that seemed to show still denser rock underlying the other layers of outer earth. Wegener called it by the name of *dunite*. Chemically, it contains iron-magnesium silicate, whereas the basaltic sima contains far less iron.

As these increasingly dense rocks are compared, they prove to be made of decreasing proportions of silicon dioxide (SO_2). Granites contain 65 to 75 percent of that compound, basalts 45 to 55 percent, while the dunite type of rock has less than 45 percent of it. Chemically, sialic rocks are described as "acid," and simatic rocks as "basic," while dunite rocks are called "ultrabasic." The latter often have a dull gray-green color, reflecting their content of a mineral called *olivine*.

The sialic rock of the continental crusts shows densities ranging from 2.67 to 2.77 times that of water. The simatic rocks of the subocean floors are denser—about 2.9 times as dense as pure water. And the ultrabasic rocks characteristic of the outer mantle are denser still—about 3.3 times the density of water. Scientists express this by saying that their specific gravity is 3.3.

Such a marked jump in density, from specific gravity of 2.9 or 3.0 to 3.3, is found where the ocean floor crust gives way to the mantle layers just below. This boundary is called, almost affectionately, "the Moho." That is short for "the Mohorovičić Discontinuity," named for the eminent seismologist Andrija Mohorovičić (1857–1936). In 1909, while analyzing data from an earthquake in Croatia (now part of Yugoslavia), he discovered, or rather deduced, the existence of this important boundary or interface.

Below the Moho, rock densities rise as much as 14 percent. Also, below the Moho, rock densities are between 20 and 25 percent greater than in the granitic sial rocks characteristic of the continental crusts.

Another boundary has a name of its own. The interface between the continental crusts and the denser ocean-floor-type crust below is sometimes called "the Conrad Discontinuity," after another earth scientist.

How far below us lie the Moho and the mantle itself? The answer depends on where we are. If we are standing on the low, level land that makes up so much of the nonocean surface of the globe, the Moho typically lies some 30 to 35 km (19 to 22 mi) below. However, under mountains there is a pronounced thickening, and the depth of the Moho increases even more than the height of the mountain itself might suggest.

Ocean floor crusts are much thinner, being roughly a quarter as thick as typical continental crusts. Hence, from 7 to 9 km (5 or 6 mi) below the sediments of the ocean floor, the Moho boundary may begin. This is the reason that subocean drilling is always proposed for the ambitious projects that seek to obtain core samples from the mantle itself, below the Moho boundary.

The distinct differences on which Wegener based his picture of continental drift processes are derived from the contrasting densities of rock layers. Thus he made his model, with the continents, like monster rock floes or rafts, being forced slowly across and through the denser sima, the strange "fluid" in which they float.

5. Continental Cross Sections and Sub-Crustal Cookery

What are typical structures under the continents and the ocean floors? A much-simplified diagram (Figure 5-1) suggests three important situations: the ocean floor case, left, at O; the ordinary low-lying dry land case at L; and the mountain case, right, at M.

At O we find five layers, each less dense than those below it. First, of course, comes the air, or atmosphere, which we need not discuss here. Next, the hydrosphere, shown as a 5 km (3.1 mi) depth of ocean water, with specific gravity of about 1.03. Next, a layer of some 3 km (1.9 mi) of accumulated sediment, its specific gravity about 2.3. Then 5 km (3.1 mi) of ocean floor crust. This is the simatic layer, in the words of Wegener, or the *lithosphere,* "zone of stone," according to modern geophysicists. Its specific gravity is about 2.9, and it terminates below, at the Moho, beyond which lies the mantle, with specific gravity of about 3.3.

The indicated depths of these layers are typical, not universal. They vary from site to site under the ocean. In particular, the depth of sediments depends on the relative age of the subocean crust below; the older the crust, the deeper the sedimentary pile collected above it. The sediment depths tend to taper off somewhat as the shoreline is approached.

A little inland from the shore, at L, where the dry land attains the typical level of about 100 m, we again find layers of increasing

Figure 5–1.

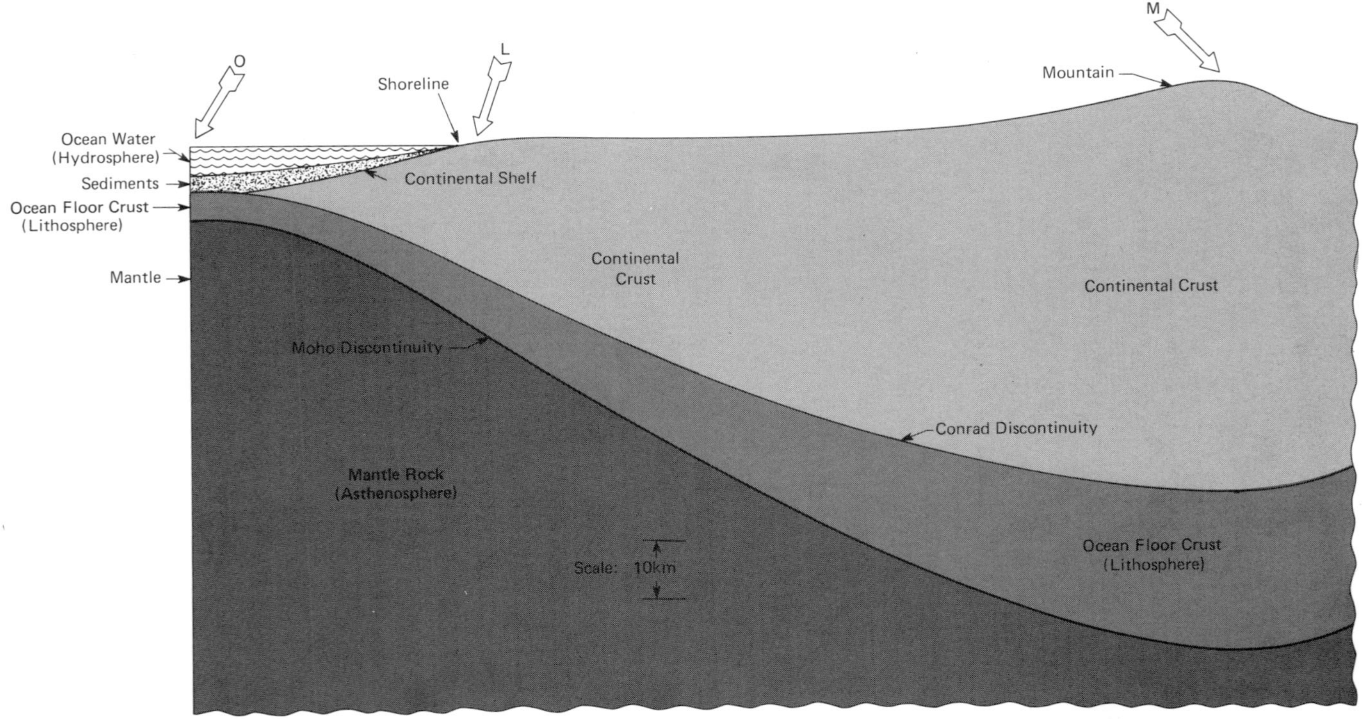
O
L
M
Shoreline
Mountain
Ocean Water
(Hydrosphere)
Sediments
Ocean Floor Crust
(Lithosphere)
Mantle
Continental Shelf
Continental
Crust
Continental Crust
Moho Discontinuity
Conrad Discontinuity
Mantle Rock
(Asthenosphere)
Scale: 10km
Ocean Floor Crust
(Lithosphere)

density—three, rather than five. First, about 20 km (13 mi) of sialic (granitelike) continental crust, on top of which may lie a thinnish layer of sedimentary soil, not shown here. This sialic crust has specific gravity of about 2.7. Below it lies about 14 km (10 mi) of the simatic or ocean floor crust, with specific gravity of about 2.9, down to the Moho boundary. There begins the ultrabasic mantle rock, specific gravity about 3.3.

At right is the mountain situation, M. The summit of the peak has been drawn to suggest an elevation of 10.1 km (33,000 ft) above sea level. Below it lies 73 km (45 mi) of the sialic continental crust, then 18 km (11 mi) of the simatic ocean floor crust. Thus, there is a total of 91 km (56 mi) of crusts on top of the Moho and the underlying mantle rock. These typical estimates are not excessive. The French geologist Jean Goguel found reason to assume a depth of some 65 km (40 mi) of continental crust under the Pelvoux Massif in his country.

These three imaginary but typical instances—O, L, and M—show much variation in total crustal thicknesses. Why should the crusts below a mountain peak have ten times or more the thickness of crusts plus sediments under a typical ocean site? We can find a clue by comparing the total masses—amounts of matter—pressing down on a particular level underground.

Let us choose the depth of 80.9 km below sea level in Figure 5-1, for the Moho boundary is shown at that depth below the mountain in M. In that case, the total mass must be proportional to the sum of two products, here rounded off to the nearest whole number:

    Continental crust: 73 km $\times$ 2.7 sp gr = 197
    Ocean floor crust: 18 km $\times$ 3.0 sp gr =   54

    Total mass units   251

In this case, the mass unit is the mass of 1 cubic kilometer of pure water, or 1 million million kilograms of mass (10^{12}kg).

Now, in the case marked by L, we must add three products:

Continental crust:	20 km × 2.7 sp gr =	54
Ocean floor crust:	14 km × 3.0 sp gr =	42
Mantle rock:	46.9 km × 3.3 sp gr =	155
	Total mass units	251

Finally, in the subocean, or O case, we must add four products for the layers of increasing density down to the depth of 80.9 km below sea level:

Ocean water:	5 km × 1.03 sp gr =	5
Sediments:	3 km × 2.3 sp gr =	7
Ocean floor crust:	5 km × 3.0 sp gr =	15
Mantle rock, down to 80.9 km:	68 km × 3.3 sp gr =	224
	Total mass units	251

Thus there is a striking and significant equality in the amount of mass or matter that presses down on a plane chosen at some suitable depth below sea level.

This kind of uniformity is found whenever less dense bodies float freely in denser substances. The equality has a scientific name, much used by geologists and geophysicists: *isostasy.* It is formed from Greek words that mean "equal" and "standing." Isostasy denotes the general equilibrium established in the crusts of the earth by the slow flow or yielding of rocky layers in response to the gravitational forces that press down on them.

If the mantle or the crustal layers above it were truly and totally rigid, isostasy could not exist. Actually, isostasy is sometimes not completely attained, because of the very slow response of rocks to continuing forces. Yet it is the prevailing *tendency* which regulates the standing or elevation of various parts of the earth's surface.

Isostasy in everyday situations can be easily demonstrated. Here (Figure 5-2) is a jar of water 10 cm deep. In it has been placed a block of wood 4 by 4 cm in cross section and 3.4 cm deep, weighing 44.6 gms. Its specific gravity accordingly is 0.82, and so it floats 82 percent below the water level and 18 percent above.

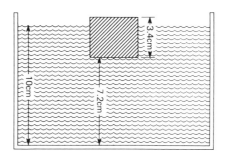

Figure 5-2.

How much mass is now supported above each square centimeter of the bottom of the jar? Again we disregard the mass of the atmosphere, which presses down equally on block and water surface. Now, excepting under the block, there are 10 cubic cm of water above each sq cm of bottom area—a mass of 10 gm.

But what about the area that lies directly under the floating block? We must add two products here:

Water:	7.2 cm × 1.0 sp gr =	7.2 gm
Wood:	3.4 cm × 0.82 sp gr =	2.8 gm
	Total mass	10.0 gm

Everywhere the equality of mass and of its downward force (weight) is maintained, when isostasy rules the relationships. We worked here with a block whose density is 82 percent of that of the water it floats in, because continental crust (sp gr 2.7) is about 82 percent as dense as mantle rock (sp gr 3.3). We could show isostasy also with blocks whose density is 89 percent that of water—the same relationship as that between the densities of ocean floor rock (sp gr 2.9) and mantle rock (sp gr 3.3).

Isostasy provides a kind of automatic compensation. Suppose we place a load of 7.8 gm on the wood block in Figure 5-2. It must now displace an additional 7.8 cc of water. Hence it sinks nearly 0.5 cm deeper in the water and floats 89 percent below and 11 percent above the water level. But if we total the average mass above each sq cm of the bottom of the jar, we find again that it is just 10.0 gm.

Thus the floating object sinks when weight is added to it and rises when weight is removed from it—and the distance of sinking or rising maintains the isostatic equality.

But what of the isostatic "floating" of entire mountains? This can be demonstrated, as in Figure 5-3, by a series of rods of various

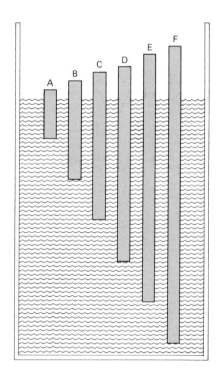

Figure 5-3.

Continental Cross Sections and Sub-Crustal Cookery 59

lengths, each with average density just 82 percent of the density of water. (Their bottoms are weighted just enough so they float upright.)

We see here that Rods A (2 cm long), B (4 cm), C (6 cm), D (8 cm), E (10 cm), and F (12 cm) each float 82 percent below and 12 percent above the water level. As they float side by side the top of Rod F corresponds to the peak of a mountain. It has below it the deepest "roots." Rod A, representing the lowland region, has far less crust underneath. Thus isostasy means that the total crust under mountains must thicken sufficiently, else the mountain peaks will sink.

What happens when mountain tops are worn away by weather? Suppose we cut off 1 cm from the top of Rod F. The rod would then rise slightly. In fact, its top would now stand only 0.18 cm below where it did before. Thus when young, jagged mountains weather and lose matter from their summits, isostasy assures that their new heights will be reduced far less than if they did not "float" in the denser rocky substances below!

On the other hand, when matter is added from outside—as by sedimentary deposits—isostasy reduces the resulting rise in altitude, for the heavier mass sinks deeper into the supporting substance below.

These effects appear everywhere on earth. For example, there has been marked subsidence (sinking) of land under many new lakes formed by dams built for hydroelectric power and flood control. The added weight of water is responsible. In the opposite direction, entire great regions of the earth, such as Scandinavia, have risen noticeably, though slowly, after the melting away of layers of ice that once covered them. Scandinavia is still rising, on its way to attaining complete isostasy.

Above all, isostasy at the surface of earth constantly demonstrates the extraordinary fact that even seemingly rigid and solid rock layers will give and flow, when subject to long-continuing forces. That too was stressed by Alfred Wegener.

Continental shelves—Figure 5-1 shows the continental shelf, sloping out a long distance beyond the coastline itself. Accumulated sediments, such as form most of these tapering shelves, are shown also. These shelf areas belong, geophysically, to the continents to which they are attached, even though they underlie varying depths of ocean water.

How far below sea level must we go in order to reach the true ocean floors, as distinct from the continental shelf slopes? Roughly speaking, something more than 1 km. This can be illuminated by an imaginary ocean drainage operation. Suppose we could lower the present sea level by 1 km. The total area of exposed "land" would then be about 8.5 percent greater than it is now. Instead of the present 29-plus percent of the total surface of earth, land would occupy nearly 38 percent. That amounts roughly to a one-third increase in land area—a sizeable gain.

But then if we could lower sea level by an additional kilometer, to 2 km below its present level, the amount of additional surface exposed would show a marked falling off. This second drainage would increase the land surface by less than one-twelfth!

The reason is that this time we would be exposing many of the steeper cliffs that plunge from the gradually tapering continental shelves down to the depths of the true ocean floors.

Roughly speaking, the earth now consists of about 60 percent true ocean floor and 40 percent continental-type surfaces. For each square unit of the continental crust, including even that part which is in the continental shelves, there are about 1½ units of ocean floor crust without continental overlay.

We have seen that the continental (sial) crusts tend to be a good deal thicker than the ocean floor (sima) crusts. It is possible that the total bulk of the former equals or even exceeds the bulk of the latter, for the world as a whole.

The following pages will show that the total area of continental crust not only does not decrease through the millions and billions of years of geologic time, but may still be increasing slowly, on the average. In the meantime, though the ocean floor areas do not increase or may even be diminishing on the whole, new ocean floor is constantly being formed and spreading in some sectors, whereas in others it is carried downward and consumed.

Losing the myth of lost continents—Alfred Wegener had sound reasons for insisting on basic differences between the crusts of the continents and those under the oceans. He laid to rest, probably for all time, the long-standing supposition that today's oceans lie on top of submerged "lost" continents of the past.

Most famous among these fabulous fictions was the legend of the lost continent of Atlantis, mentioned in one of the subtle dialogues written by the famous Greek philospher, Plato.

Wegener rejected the notion of the sunken continents. Instead, he showed that continents are formed one way, ocean floors another. The former never become the latter, nor vice versa.

Continents do move, Wegener told his readers—but they move "horizontally," over the face of the globe. They do not move downward vertically to become the floors of some future ocean. Nor does true ocean floor rise up above sea level and appear in the form of a new continent. True, there have been long periods in which great areas of continents lay under water—but this water only formed shallow seas, not true oceans. One might almost say that an ocean's an ocean, a continent a continent, and never the twain shall exchange their roles on earth.

Yet even while Wegener insisted on the fundamental differences between the sial of the continents and the sima of the ocean floor crust, he noted a great likeness between the two types of rock. Both are what geologists call *igneous* and *magmatic. Igneous* means originating under conditions of great heat, commonly at temperatures high enough to melt the materials. *Magmatic* means formed from *magma,* the molten material forced up from the interior to the surface under conditions commonly called volcanic (Figure 5-4).

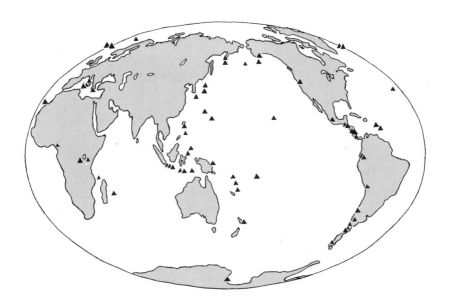

Figure 5-4. Principal groups of active volcanoes, indicated by black triangles. More than 60 percent of them are found in the great "ring of fire" around the Pacific Ocean.

Along the Pacific coast of North America the volcanoes form a well-defined chain. Volcanoes are classed as active, quiescent, dormant, or extinct. Many North American volcanoes are commonly considered to be extinct, but geologists specializing in the field believe that sooner or later each of them may erupt again.

The California volcanic chain includes, from south to north, Mounts

Lassen and Shasta; in Oregon-Washington, Mounts Hood, St. Helena, Adams, and Rainier; also Glacier Peak and Mount Baker.

The volcanic and thermal activity in this zone is linked to the seismic or earthquake activity, thanks to the insights of continental drift theory.

Based on data from K. Sapper.

Both the basalts of the ocean floors and the granites of the continental crusts were originally "cooked" deep down below, in what is now the mantle of earth. Then they floated—or were forced—upward and cooled to their present forms.

The earth still operates like an enormous pressure cooker. In well-defined volcanic regions, molten magma and lava emerge at the surface, sometimes explosively, sometimes as slow flows. This molten raw material becomes rock of varying degrees of density and porosity, depending on the conditions of pressure and temperature prevailing when it solidifies.

Sometimes the newly baked rock is as light and porous as pumice. Sometimes it is as dense and tough-grained as the basalt sheets of ocean floor crust which are constantly emerging from the depths, as we shall see.

These titanic cookery processes resemble what happens in smelting furnaces as crude ores are melted. The heavier metals sink toward the bottom and the lighter slags rise upward. On a homelier scale, much the same happens when a housewife cooks a soup or boils a mixture of fruit, sugar, and water to make jam.

The continental crustal rock (sial) is like the lighter slag or froth that emerges from the great geological cook-outs in the earth. It may seem strange to equate hard granites with froths, but the analogy is justified.

The simatic rock of the ocean floors, though markedly denser than the sialic rock of the continental crusts, is still less dense than the ultrabasic rock of the mantle layers below. Thus the oceanic crust can be assumed also to have cooked up from the hot mixtures that still remain below the Moho boundary.

The great physical differences that mark the surface of the earth—the continents, islands, oceans, and all—have been produced from a supply of ingredients or chemical elements that do not differ nearly as much as one might expect. It is like meals of the utmost variety and novelty prepared by a skilled cook from one kitchen and one larder stocked with a rather modest supply of ingredients.

About 99 percent of the earth's solid crusts are composed of just eight chemical elements, given here with rounded percentages: oxygen (47 percent); silicon (28 percent); aluminum (8 percent); iron (5 percent); calcium (4 percent); sodium (3 percent); potassium (2½ percent); magnesium (2 percent).

Several other elements amount to only a fraction of 1 percent, as for example titanium (0.5 percent), phosphorus (0.12 percent), and manganese (0.1 percent). The remaining elements each constitute less than 1 part per 1,000 of the earth's crust. The ingredients essential to living creatures are actually extremely *rare* in the great larder of the earth's crust. Thus, carbon is present only to the extent of some 3 parts per 10,000, and sulfur 5 parts per 10,000.

Some of the substances basic to human technologies are among the rarest. Copper, for example, amounts to less than 1 part in 10,000, lead to less than 2 parts in 100,000. Silver and gold, which have ruled the lives of so many on earth, amount to less than 1 part in 1 million and 1 part in 10 million, respectively!

However, in the long list of some ninety elements which are sparsely present in earth's crust, we find two whose enormous importance seems the more striking because of their comparatively extreme scarcity: thorium (about 1 part in 10,000) and uranium (less than 1 part in 100,000).

Was the earth once molten?—The heat sources as well as the ingredients of the great kitchen of earth are important for geophysics. When Wegener began his work on continental drift, the general belief was that the earth had originated as a molten mass. Today, on the basis of more complete information and analyses, it seems likely that never in its 4 or 5 billion years of existence was the earth entirely or even largely molten.

However, it was hot enough inside to make possible extensive separations and slow circulations of substances. The present multilayered structure of the earth strongly suggests this.

We have seen that the lighter "froth," high on the surface, corresponds to the continental plates; a somewhat denser layer corresponds to the ocean floor crusts of basalt (sima); and under both, past the decisive boundary of the Moho, we find the great zone of the mantle, formed of rock even denser than that in the outer crusts.

The earth itself has an average radius of some 6,370 km (3,960 mi). Does the mantle occupy all this enormous interior? By no means! Just as the exterior crusts give way to the thick mantle at the Moho, so at about 2,870 km below the surface the mantle gives way to the remarkable core.

That core, composed of iron and some nickel, now has a radius equal to about 55 percent of the total radius of the earth. The core contains, in fact, about 17 percent of the earth's total volume.

Even the core is divided into layers: an inner core of solid iron and an outer concentric layer that appears to be liquid iron, despite the enormous pressures at that depth.

At ordinary atmospheric pressure on earth's surface, iron has a density about 7.8 times that of water. Under the huge internal pressures, iron in the earth's central core is compressed to 11 or 12 times the normal density of water. The large, dense core accounts for the fact that the earth as a whole has a notably high density. Its average density is, in fact, more than 5½ times that of water. Among the other planets only little Mercury shows higher average density, being less than 2 percent denser than our earth.

The dense, heavy iron core at earth's center matches the model mentioned earlier—the smelting furnace or the boiling soup kettle. In each, the heat-powered flow of substances is separated into ingredients that gather in different places according to their different densities. The least dense rise to the surface and float there; the densest drop to the bottom and stay there. The "bottom" in our earth is, gravitationally speaking, its center.

Alfred Wegener's book, and some insights into earth's structure attained since his time, allow us to conclude that the same processes that surfaced the stone "froth" of the continental crusts also massed in the center the iron and nickel which form the formidable heart of the planet—its core.

But just *how* might such enormous separations have been started and carried through to the present stage?

They could not have been completed while the earth was still quite young, for the very oldest rocks that have been found on the continental surfaces of earth do not date back to more than about 3.6 billion years ago—a time when the earth was already roughly 1 billion years old. In fact, geologists commonly describe as "old" formations such as some of those which make up the stone skeleton of Africa; yet those date back less than three-quarters of a billion years, to a time when the earth was already about 4 billion years old.

By far the majority of the continental rocks that can be studied by the surface-limited scientists of today have been formed during the latest third or quarter of the total age of the earth. And the rocks of the ocean floor crusts are decidedly and astoundingly younger still —all of them!

Truly, the evolution of today's earth, however it took place, has been a long-continuing process. Even a couple of hundred million years ago—when 95 percent of the earth's present age already lay behind it—the surface looked far different than it does now. And there is abundant reason to predict that the earth of 50 or 100 million years from now will show its inhabitants a far different face—if, indeed, human inhabitants then remain alive to look upon it.

6. The Mountain-Making Mother . . .

The title of this chapter is taken from one of the poems of Robinson Jeffers, already mentioned for his understanding of modern scientific ideas. The mother is, of course, our Mother Earth, who has raised up so many imposing mountains in the past, and has not yet ceased to make more.

Alfred Wegener, in the second chapter of his classic work on continental drift, dealt with mountain-making. He offered interesting proof of the closely matching characteristics of mountains on continents now far from where they presumably were placed when the mountains were formed.

Thus Wegener showed detailed correspondences between parts of western Africa and eastern South America. Complex sequences of stratified rock are duplicated or closely paralleled in both sites. Even the South African regions where white diamonds (crystallized carbon) are found in so-called pipes have counterparts in the Brazilian state of Minas Gerais.

Various geologists have listed such geological matchings in much detail. Especially persuasive was the work of Alexander du Toit, a South African geologist. In writings published during the 1920s and 1930s, he strongly supported continental drift ideas. Wegener was especially impressed by du Toit's 1927 study, *A Geological Comparison of S. America with S. Africa.* Wegener said that until he had read it he had "hardly dared to expect so close a correspondence between the two continents."

The Madagascar-India match—Wegener also stressed other comparable matches between widely separated land areas, including India, whose west coast, he concluded, had once joined the east coast of Madagascar. The enormous compressions that created the Himalaya Mountains must have been produced as India, parting company with Madagascar, moved northward, pushing at last into and under the plate of southern Asia itself.

Like a geophysical Magellan, Wegener circumnavigated the world, assembling evidence for continental drift from the labors of many geologists. Though the drift theory was not supported by the majority of earth scientists in the 1920s, Wegener reported that it already had shown itself a useful tool for professional geologists. For example, various Dutch geologists grappling with problems of the regions south and east of Indonesia (then a Dutch colony) had been among the first to accept and apply assumptions of drift theory. One of them, G. L. Smit Sibirga, stated in a 1927 paper that the theory provided acceptable answers for a dozen different problems that previously had stumped geologists.

Wegener quoted also a French scientist, E. Argand, who in the course of a 1924 study of the tectonics of Asia had informed his colleagues that "the theory of drift of the large continental masses" enjoyed "an excellent state of health." He described it as being based firmly "on the overlapping areas of geophysics, geology, biogeography, and paleoclimatology."

The life sciences contribute too—Alfred Wegener included evidence and arguments from all these sciences in his book. He compared past distributions of animals, as determined by paleontology, with their present distributions, described by zoogeography. The duplications between distinctive reptiles and mammals in western Africa, on one hand, and South America on the other, indicated that until some time *after* those species had evolved, the two great land masses must have been joined.

In the North Atlantic regions this situation was less clearcut. However, even here evidence for drift was found in the distributions of animals as varied as perch (fish), mussels, snails, and earthworms; also in the distributions of some plants. Unless one assumed drift, one had to assume the former existence of an enormous land bridge, now vanished. But vanished *to where?* To that question no one could supply a satisfactory solution.

Identical or very similar plants and animals in India and Madagascar provided evidence that both regions had once been joined together. Similar evidence, faunal and floral alike, linked Australia with parts of South America, on one hand, and with New Zealand on the other.

Thus Wegener traced the great link-ups that had been ended by tearings apart and subsequent drifts.

Climate as a climactic argument—In the seventh chapter of his book, Wegener offered arguments from his own special fields: climate in general and ancient climates in particular. Fossil remains or imprints of plants and animals indicated that great land masses once must have had far different positions than they do today with respect to the earth's axis of rotation.

Wegener pointed out repeatedly still another kind of major migration or drift—the so-called *polar wandering.* The earth's axis of spin was not always where it is now, between the present North and South poles. These axial shifts were separate from the motions of individual continents, but had to be taken into account if the latter were to be understood properly.

When both the axial and the continental wanderings were sorted out, great clarification resulted. Said Wegener, "The former confusion of disorderly and seemingly self-contradictory facts link up to form a pattern whose simplicity repeatedly astonishes me. . . . ." Wegener devoted a chapter to this sorting-out.

Some of his findings are shown here (Figure 6-1). At left is the way the South Pole had moved, as seen from South America; at right, the way it moved, seen from Africa. Each of these "itineraries" of the earth's spin axis is based on four different geological periods. The earliest, marked "Cret." for Cretaceous, began some 130 million years ago, though in Wegener's time it was given a more recent date. Next was the Eocene, which Wegener dated at about 15 million years ago; then the Miocene, about 6 million years ago; and finally the Quaternary, about 1 million years ago.

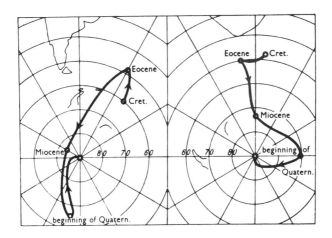

Figure 6-1. Past wanderings of the South Pole as seen from two different and significant points of view: left, from South America; right, from Africa.
Reproduced by permission of Dover Publications, Inc.

These numbers of years are only approximate. The important aspect here is the *extent* of the shifting of the axis of rotation, indicated by the circles of latitude surrounding the present South Pole. Whether seen from Africa or South America, that shifting covered more than 30° of latitude, or about 3,500 km (2,170 mi) distance from today's South Pole.

As seen from the South American point of view (at left), the southern end of the axis of rotation must have moved across more than 50° of latitude between the Eocene and early Quaternary periods. If this motion were laid out along a single meridian of earth, it would reach from the present South Pole to a point north of the equator!

From the African viewpoint the polar shifts appear not quite so extensive. The reason is simple: Africa shifted less, with respect to the present position of the South Pole, than did South America.

Cities that shifted—During the combined polar and continental shifts, the latitudes—and hence the climates—of important sites on earth changed drastically. For the well-known Saxon city of Leipzig, now at about 51° N latitude, or about 57 percent of the way from the equator toward the North Pole, Wegener presented this itinerary of past travels: Leipzig was at the equator when the coal beds were laid down in the Carboniferous era. By the Permian period it had moved to 13° north, and to 20° north in the Triassic. By the Miocene it was at 39° north, and by the beginning of the Quaternary it was at 53° north, closer to the pole than it is even now.

Europe in general moved northward, and so increased its latitudes, between the Carboniferous and the close of the Triassic period. Then came an interval of decreasing latitudes (southward motion), and finally a period of substantial northward movement again.

While Europe drifted southward relative to the earth's rotational axis, vast changes took place in it—especially the formation of huge, shallow inland seas during the Jurassic and Cretaceous periods.

Large segments of Europe were quite covered by such sheets of water, which were by no means true oceans.

Our unsolid earth—No wonder Wegener, in the face of all these complex changes, concluded that "the earth is not a solid body," but rather one that "exhibits flow and is subject to continental drift, crustal wandering, and probably also to internal axial displacements."

An earth which thus "exhibits flow" is not necessarily the same as a fully *liquid* earth. Many substances behave like crystalline solids, and are even rigidly brittle, when subject to sudden, sharp forces—but will slowly creep or deform when long-continuing pressures or tensions are applied. Glass is such a substance. Even more surprising is the substance sold as an amusement under such names as Nutty Putty and Silly Putty. It can be bounced like a ball or shattered by a hammer blow; yet when set down for a long time, its own weight suffices to make it sag and flow like some thick liquid.

Wegener's evidence, and further researches that followed his work, reveal our earth to be neither fully fixed in shape and structure, nor finally formed. Instead, one could recall the prophetic final statement by the magician Prospero in Shakespeare's *The Tempest:*

> The cloud-capped towers, the gorgeous palaces,
> The solemn temples, the great globe itself,
> Yea, all which it inherit, shall dissolve
> And . . .
> Leave not a rack behind.

Strange but unmistakable records in the rocks of earth show that the former states of our great globe have changed and the present states are even now changing further.

What drives the drifting continents?—In the ninth chapter of his historic book, Wegener tried to come to grips with a most difficult problem: What forces have moved entire continents such distances? As a principal answer he proposed a source of power and a mechanism that is almost certainly *not* the right one. Therefore it will be but briefly presented here.

Look again at Figure 5-3. Where is the center of gravity of floating rod F? It is halfway down the 12 cm rod, hence 6 cm below its top, and about 3.4 cm below the level of the water. But where was the center of gravity of the column of water which that floating rod displaces? That imaginary column of water was 9.8 cm deep; hence its center of gravity was 4.9 cm below the surface—or about 1.5 cm farther down than the center of gravity of the entire rod. Thus, the floating rod's center of gravity is *higher than* the center of gravity of the liquid it displaced.

In the same way, each continental block or plate has a higher center of gravity than the substances (simatic rock of ocean floor crust or ultrabasic rock of the mantle) that it displaces.

The earth spins constantly on its axis, once in twenty-four hours. As with any merry-go-round or revolving platform, this spin produces centrifugal force, tending to move objects outward, away from the axis of rotation. That axis is the imaginary line from pole to pole, passing through the center of the earth. In fact, a man standing on the equator weighs about 1 part in 290 less than he would standing on one of the poles, simply because the centrifugal force nullifies some of his weight.

The earth's spin tends to move—away from the poles and toward the equator—massive floating objects. The reason is that their centers of gravity are higher than the centers of gravity of the supporting material that they displace. The equator is about 6,370 km *out* from the axis. Hence a move from a pole toward the equator is a move outward, away from the axis of rotation.

Wegener called this small tendency a "pole-fleeing force." He could also have called it an "equator-approaching force." He did not claim to have discovered this force. It had been described as far back as 1913 by a distinguished Hungarian geophysicist, Baron Roland Eötvös (1848–1919), who had noted also that the same kind of force acted oppositely on portions of the earth's outer layers whose centers of gravity were *lower* (closer to earth's center) than the centers of gravity of the substances they displaced. Such bodies would tend to be drawn, or forced, toward the poles and away from the equator.

Earth's floating continents assuredly have higher centers of gravity than the denser rocks in which, or on which, they float. Hence, Wegener argued, continents in the Northern Hemisphere should tend to move southward, and those in the Southern Hemisphere northward. However, the Atlantic Ocean was born mainly from *westward* movement by the continents of North and South America. Wegener sought to explain this seeming contradiction by recalling that the "pole-fleeing force" had worked during millions of years when the earth's axis was far differently positioned than it is now.

Many a critic of Wegener's work has complained that he failed to offer a plausible source of the huge power needed for continental drift processes. Actually, the Eötvös "pole-fleeing force" could exert on a continent a force amounting only to a few parts per million of the force represented by the (downward) weight of that continent.

Wegener, however, paid attention also to other possible sources of power. In the last edition of his book he wrote that "a circulation of sima beneath the crust" seemed to offer "a reasonable explanation" for the great continental drifts that opened up the Atlantic Ocean. Here he put his finger on processes that a great many leading geophysicists in the 1970s believe must move continents, generate earthquakes, and maintain volcanic activity in the surface layers of our earth.

It is the mantle that lies "beneath the crust," and the "circulation of sima" to which Wegener referred is commonly called heat convection currents in the mantle. Such monster convection currents will be pictured in closer detail in following pages.

Clearer effects than causes—Though Wegener, from his vantage point of the 1920s, could not solve the problems of the source of power for drift and related effects, he had no doubt that the necessary forces *were* at work, and he was clear as to the range of results that they produced. He stated specifically that "the forces which displace continents are *the same* as those which produce great fold-mountain-ranges."

The connection between drift and mountain-making seemed the more clear because both had occurred together and only during "certain periods of the earth's history." There were episodes without continental drift, and without mountain-making, too.

Yet Wegener tried repeatedly not to become too dogmatic. He cautioned his readers—and perhaps himself—that "what is cause and what is effect, only the future will unveil."

Today, nearly half a century after this persistent pioneer died in Greenland, that unveiling has gone a long way further, though it still has far to go. We may even risk a forecast: By the year 2000, the causes and the effects of these great surface changes on the earth will have become clear in their major outlines, though the details will be far from complete.

The new insights called *global tectonics* are already reducing the uncertainties. The progress of the years since 1960 has been in many ways enormous, and the tempo seems to be increasing.

7. Bumping Blocks and Quaking Crusts

Alfred Wegener concluded the final edition of his book with chapters intended to raise questions and stimulate discussion. He noted that the continental crusts of sialic rock are largely overlaid by deposits of sediments, ranging in depth from a few meters to as much as 10 km (6.2 mi), but averaging about 2.4 km (1.7 mi) thick.

In all, about 60 km (37.2 mi) separate the top surface from the ocean floor crust of basalt or sima underneath. Of this distance, the top 4 or 5 percent is likely to be sediments, including soil; the next 45 or 46 percent is made up of various kinds of nonigneous rock; and the lowest 50 percent is the granitic sial crust peculiar to continents.

This is the first mention in this book of rocks which are *not* igneous, as are the granites, the basalts, and the ultrabasic rock of the underlying mantle. Two major kinds of rock make up the nonigneous category, *sedimentary* rock and *metamorphic* rock.

Sedimentary rocks originate from showers of particles or sediments, often deposited under water or even laid down by winds, then compacted and consolidated. Such common substances as soil, gravel, and clay are sedimentary in origin. So, too, are the myriads of remains of marine animals or plants that form chalk deposits. When such calcareous matter is firmly cemented together it becomes the sedimentary rock called *limestone.* Sedimentary grains of quartz, similarly cemented, form sandstone. Most of the strata under England, for example, are various kinds of sedimentary rock.

Metamorphic means "changed" or "transformed." Rocks in this group begin as either igneous or sedimentary substances. Then, as a result of extremes of temperature, pressure, or chemical processes, they are altered in structure, texture, and even in composition. Slate, mica, marble, gneiss, schist, quartzite, anthracite (coal), and graphite are all metamorphic in origin.

In general, and apart from special local situations due to major earth movements or upsets, the sedimentary rocks tend to be found above the metamorphic and the metamorphic above the igneous. Material may alter or evolve from one rock form to another. Thus granitic (sial) rock, rich in silicates, may disintegrate into sandlike bits, which then are deposited under water as sediments, compacted into sandstone, and later metamorphosed by pressure and heat into quartzite. Or sedimentary clay may become rocklike shale and later be metamorphosed into slate or schist. Organic plant remnants, after decaying to become peat, may be transformed into sedimentary lignite or bituminous coal and later be metamorphosed into anthracite or graphite.

The enormous mile-deep gash in the earth called the Grand Canyon reveals layer after layer of sedimentary rock, until near the bottom where the swift river has cut through to the still lower metamorphic rocks.

True igneous rocks are not layered. However, they may form faults and cracks. Some igneous rocks, formed of magma that has been very quickly cooled from its molten state, become quite glassy in structure—like obsidian or volcanic glass. Even such solid, smooth rock, however, is eventually attacked and worn down by the actions of weather, running water, and plant life at the surface. In frigid climates, frost splits rocks and glaciers grind them.

Thus, through long ages, sediments accumulate on continental land surfaces and on continental shelves or platforms below sea level. These sedimentary collections may be pushed together until they emerge above sea level, forming new dry land. Such processes help not only to maintain but even to increase slowly the total land areas of the globe as a whole.

Basic blocks beneath—Under all the surface materials and layers that may have gathered on the continents and the ocean floors lie the great supporting blocks or plates of igneous rock. The continental plates ride, or float, on the ocean floor plates, which in turn rest on the great mantle layer below. The boundary just above the mantle is the Moho.

Continental drift can take place only if such enormous plates move slowly but substantially with respect to each other. Wegener described the principal ways in which such motion could occur, and showed what surface features should result from each kind of relative motion.

Here, in Figure 7-1A, is the simple case of two plates, X and Y, in head-on collision. The effect is the same whether X moves toward Y, Y toward X, or both toward each other. Total motion, we must remember, may be no more than 3 to 15 cm (1.2 to 6 in) per year.

The result of the motion shown here is compression, producing folding of sediment and rock in lines lying parallel to the boundary between X and Y. Wegener recognized folding like this in mountain systems such as the Andes of South America.

Today's tectonics regards the typical result of such "head-on collisions" not as the accordionlike folding of both plates, but rather as the overriding of one plate (the continental plate) and the underthrusting of the other (the ocean floor plate). Thus, if plate Y carries the continent and X the ocean floor, the latter is typically forced under the former. The continental areas remain enduringly on the surface, safe against destruction. The ocean floor, however, is carried down until it melts and merges in the mantle mass below.

Thus it is on the overriding plate that the folds (mountains) parallel to the boundary appear.

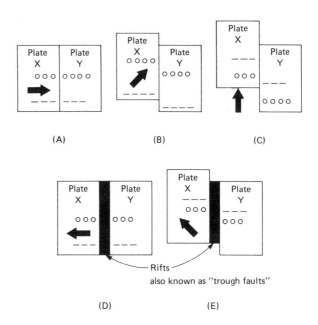

Figure 7-1. Five fundamental kinds of encounter between adjacent plates, as described by Wegener. The dark arrow indicates the relative motion of Plate X (at left) with respect to Plate Y (right). In reality, Plate Y is seldom fixed or motionless. All the great tectonic plates of the earth are found to be in motion, relative to one another.

Angled approach—Plate X may be moving toward Y but at an angle, as in Figure 7-1B. Wegener noted that the result is then staggered folds, "which are closer and lower, the more [nearly] the direction of movement is parallel to the boundary between the blocks."

The relative motion of the plates may even be parallel to the boundary between them, as in Figure 7-1C. The result here is a great fault-forming system. The boundary becomes what geologists and seismologists call a "strike-slip fault." Valleys, ridges, or roads which once ran straight across the fault line are now offset. Each new strike-slip shift increases the offset or staggered effect, as suggested by the dashed and dotted lines on the diagram.

The plane of the fault may slant down deep below the surface. Such fault planes are seldom "straight down" in direction. On the whole, this situation is identified with displacements and staggered features on the surface, rather than with folds ridging the surface.

What if the direction of motion is exactly opposite to that in Figure 7-1A? In other words, plates Y and X are pulled *away from* each other, as in Figure 7-1D. As they pull apart, a gap or rift is left between them. The result is an ordinary rift. This rift runs at right angles to the direction of the motion that produced it.

However, there are innumerable possible variations in the direction of such separating motions. Figure 7-1E shows plate Y moving away from X at an angle, not perpendicular to the boundary between Y and X. The earmark of such oblique withdrawal, as Wegener noted, is the formation of oblique or angled rifting.

Rifts of all kinds are common on earth, both above and below sea level. Great systems of so-called rift valleys, some covered with shallow water (lakes), exist in East Africa and elsewhere. The Red Sea itself is a continuing and spreading rift. Rifting is even now enlarging the Gulf of California, a rift formation.

Modern tectonics finds that when two plates form rifts by their pulling apart, the resulting thinning or splitting may finally open gaps through which molten magmatic material from deep down in the mantle oozes upward, then hardens after entirely or partially filling the rift space.

The last stages of such rifting prior to the emergence of magmatic material from the mantle are probably found today at the so-called Afar Triangle, a strange depression at the southern end of the Red Sea. And the first formation of new, thinner crusts between the receding plates has apparently taken place in the northern Red Sea, between Africa and Arabia.

Once, some 180 to 200 million years ago, such a rifting process must have first opened the gap between Eurasia and North America —the gap that expanded finally to become the northern Atlantic Ocean. Similar rifting moved South America from Africa to form the South Atlantic. In each case, what had been a single continuous or fused tectonic plate was strained, split, rifted, and separated.

Strike-slippage close to home—Wegener supplied an example of a strike-slip fault on a grand scale. He called it the "San Francisco earthquake fault," though it is better and more correctly known now as the San Andreas fault system. Here, in Figure 7-2, are his maps showing relative motions of the great plates, for California as a whole (left) and on a larger scale (right) for the region from just south of San Francisco Bay to north of Cape Fortune.

This great fault line today is traced much further south than Wegener showed it. The line is now known to reach into and through the gulf which separates Baja (Lower) California from the Mexican mainland.

The writer of this book, residing and working on the Pacific Ocean shore somewhat west and north of Los Angeles, is especially aware of the San Andreas fault system and its possibilities or perils. It lies not many miles inland!

The bulk of California, including San Francisco itself, lies east of the San Andreas fault and consequently seems constantly to be shifting southward, with respect to the land west of the fault system. Hence the writer's chair, typewriter, and desk, as well as the sprawling, smog-choked metropolis of Los Angeles, and so much else—all appear to be shifting or jumping northward, as seen from the other side of the fault structure.

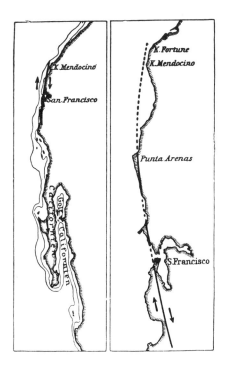

Figure 7-2. A famous fault, site of incessant earthquakes, small, medium, and large. At left is Wegener's diagram showing what was then known of the San Andreas fault system, especially north and south of San Francisco. At right, an enlargement of one section, showing how the fault splits the peninsula at the northern end of which San Francisco is located.
Reproduced by permission of Dover Publications, Inc.

Here, at continent's end, we have a prime instance of two great tectonic plates in contact, their margin forming fault structures like those in Figure 7-1C.

Enormous friction tends to lock together the edges of the American plate to the east and the Pacific plate to the west of the fault structure. But the relative motion continues inexorably, the mighty rock layers are strained out of shape, and then—suddenly—comes one of the great jumps that human beings experience as earthquakes.

Such a release of stored-up energy took place April 18, 1906, along the fault through the San Francisco peninsula. Result: the catastrophic event called the San Francisco Earthquake.

Under the coastal regions of California and other active seismic zones, the earth is repeatedly shaken. Its irregular motions and

Figure 7-3. The epicenters of the world's earthquakes outline the great tectonic plates of our globe. Each black dot on this map shows the epicenter of an earthquake that occurred during 1961-67.

Heavy concentrations of quakes appear as broad black bands, like those outlining the "ring of fire" around the Pacific. These mark the deep-ocean trenches into which ocean floor crust is thrust and swallowed up. Here the earthquakes are not only most frequent, but also commonly deeper in location, occurring as far as 700 km below the surface. The lighter lines, like the curves tracing out the Mid-Atlantic Rise, mark the rise and rift systems. Here ocean floor crust is pulled apart and new crust emerges from below, filling the gap.

West of the continent of South America, this map shows a loop indicating the margins of the East Pacific plate, which is being moved toward that continent and pushed into the trenches along its western edge. Also shown are the strangely curving plate outlines that take in both Australia, the island continent, and India, the great subcontinent that has been pushed into and partly under the Eurasian plate north of it.

Earthquake data from ESSA (Environmental Sciences Services Administration, U.S. Department of Commerce). Map reproduced by courtesy of M. Barazangi, J. Dorman, and the *Bulletin* of the Seismological Society of America.

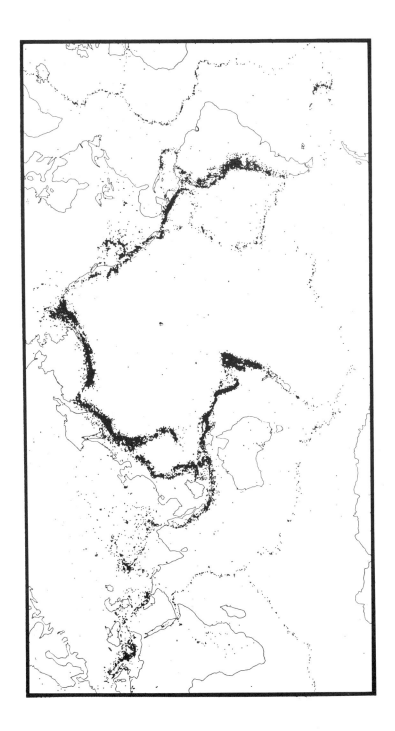

vibrations are given various names, such as earthquakes, temblors, tremors, aftershocks, and microseisms. There is not a day or hour when the earth is not quaking somewhere. Beyond doubt, quakes of a magnitude as great as or greater than any previously recorded will take place in the future. Just when, however, no one can say. Seismic science and geophysics are not yet advanced enough to predict precisely when or where the next will strike. Yet rapid strides are being made toward techniques that may soon make some forecasting possible.

Wegener, writing in the late 1920s, was clear as to what went where to make the great San Francisco quake: "The eastern section moved suddenly southward and the western section northward." He knew also that the relative jump was greatest above the focus or center of the earthquake, and that the observed "extent of the sudden shift decreased with distance from the rift and at long distances could not be detected."

Not only the Pacific coast is active earthquake country. So, too, is virtually the entire rim or margin of the Pacific Ocean, a zone sometimes called "the ring of fire" (Figure 7-3).

Wegener also considered strike-slip faults in the coastal regions at the west of Indochina. He knew that major earthquakes are concentrated in well-defined and closely connected zones or bands, though he could not know that these bands themselves constitute a clue to the outlines of the major tectonic plates of the earth.

Shaking clues to the earth's interior—In recent years, seismologists have been able to locate earthquake centers, not only with respect to positions on the earth's surface above them, but also in terms of their distance downward below the surface. Such precise placement of the sudden shifts has added enormously to the understanding of processes in and under the crusts that cover the earth.

The origins of earthquakes vary widely. The very deepest recorded have been about 700 km (435 mi) down, or more than one-sixth of the way to the bottom of the mantle itself. A larger group lie between about 300 and 650 km (190 to 270 mi) down. And in extensive zones, almost all the earthquakes are shallower than 300 km. Highly significant records reveal that earthquake locations on the continental side of deep-ocean trenches show increasing depth with increasing distance from the low point of the trench—just as if the quakes took place along a sort of inclined plane originating in or near the trench itself.

Distant earthquakes are detected and placed by means of the earth-shaking waves that travel from them to instruments around the world. Wegener in his book offered a map of the Pacific area (Figure 7-4) showing the focuses of earthquakes that had been recorded and calculated in Hamburg, Germany. The small crosses locate earthquakes whose waves as received at Hamburg had long periods, indicating that they had traveled through the denser ocean floor crusts of the Pacific, the Atlantic, or the North Sea. The black dots, on the other hand, locate earthquakes whose shorter periods resulted from their having traveled through the less dense continental crusts, via the Eurasian plate.

The actual speed of wave travel had differed, being about 3.7 km (2.3 mi) per second through continental crust, but some 18 to 22 percent faster through the denser ocean floor crust. Thus seismic records, too, helped to prove that "the ocean floors are made from fundamentally different material" from that of the continental blocks.

Much as the electromagnetic waves called X-rays are used to see through many solid objects, so the seismic waves have enabled geophysicists to "see" surprising details within the earth. Our knowledge of the central iron cores, for instance, is based entirely on seismic records and analyses.

Earthquakes, volcanic eruptions, thermal upheavals such as gey-

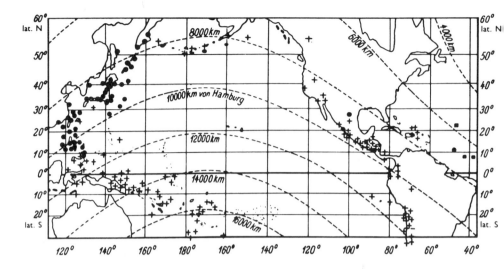

Figure 7–4. Earthquakes around the Pacific "ring of fire," whose waves traveled to Hamburg, Germany, disclosing differences in the crusts through which they passed.

The dashed curves indicate distances the waves traveled, ranging from 16,000 km (9,900 mi) at the bottom, to 4,000 km (2,480 mi) at upper right. Crosses locate the earthquakes whose after-waves reached Hamburg through ocean floor crusts. Fat black dots show those whose after-waves went through Asiatic continental crusts. The lower densities of the latter resulted in shorter periods of those waves as recorded on the instruments at Hamburg.

Reproduced by permission of Dover Publications, Inc.

sers and hot springs—all these more violent evidences of the internal strains and high-temperature torments of the earth have supplied information on which present global tectonic concepts are based.

A quiet, static earth might be closer to the preferences of some of its human inhabitants. But it would reveal far fewer of its secrets to scientists.

The dynamic origins of great earthquakes, as well as some of their effects on people who experience them, have been described by the

Bumping Blocks and Quaking Crusts

poet Robinson Jeffers, already mentioned for his imaginative insights into science. He pictured in this way the stored-up forces hidden under the mountains of the California coast on which he lived:

> Not for quietness, not peace;
> They are moved in their times. Not for repose; they are
>   more strained than the mind of a man; tortured and
>   twisted
> Layer under layer like tetanus, like the muscles of a
>   mountain bear that has gorged the strychnine
> With the meat bait: but under their dead agonies, under the
>   nightmare pressure, the living mountain
> Dreams exaltation; in the scoriac shell, granites and basalts,
>   the reptile force in the continent of rock
> Pushing against the pit of the ocean, unbearable strains and
>   weights, inveterate resistances, dreams westward
> The continent, skyward the mountain . . .

The westward, or rather northward, thrust of the coastal strip on the ocean side of the great fault achieves a sudden release:

> The old fault
> In the steep scarp under the waves
> Melted at the deep edge, the teeth of the fracture
> Gnashed together, snapping on each other; the power of the
>   earth drank
> Their pang of unendurable release and the old resistances
> Locked. The long coast was shaken like a leaf . . .

In this poetic novel, *The Women at Point Sur* (1927), one of the characters, experiencing the consequences of this seismic shift,

> felt
> The floor under her feet heave and be quieted.

Others, not far away, hear

> the sparrows
> Cry out in the oak by the window, in the leonine roar
> Of the strained earth, the clatter of bricks or small stones,
> And the great timbers of the walls grind on their bearings.

Likewise,
> the people
> Camping beside the creek under the mountain
> Heard the hills move, heard the woods heaving, the boughs
>   of the redwoods
> Beating against each other from the southwest,
> And the roaring earth. The earth swayed in waves like a
>   bog.

Another character elsewhere
> heard the forest and the stones falling
> And the roaring earth. He felt the limestone mountain
>   shaken
> For a willow leaf in a light wind by the streamside.
> When the mountain was quiet his body had ceased
>   trembling.

The extreme emotional shock that often comes with great earthquakes and the accompanying need for human contact and comfort are suggested in a brief poem, "Earthquake," by a high school student, Holly Zimmerman, of Los Angeles:

> People brought together
> By something that tore the earth apart.

In a later poem, "Resurrection" (1933), Robinson Jeffers describes how, when "one of the local earthquakes/Moved in the mountain," a woman along the coast

> heard the thunder of the hidden rock and felt the soil of

> the hillside quiver; when she looked up
> The little house seemed to be dancing by itself, the dark
>   mane of oaks
> On the spine of the hill nodded above; then instantly she felt
>   the whole mountain lifted a little,
> And waver and fall, as if some power in the rock weighed it
>   on bloody shoulders, struggling to rise.

Later in this book we shall look more closely at "some power in the rock" sufficient to account for the motions that make earthquakes, that fold and form mountains, and that move great continents long distances over the face of the earth.

8. Convections, Conductions, and Temperature Differences

Alfred Wegener emphasized the large-scale and long-time connection between five different kinds of occurrences on earth: (1) continental drift, (2) cycles of spreading and dwindling of inland seas, (3) huge faults and compressions found on the earth's surface, (4) earthquakes, and (5) volcanic action.

All five of these seemingly separate kinds of events showed marked increase during certain periods. This concerted "on and off" timing suggested strongly to him that common causes underlay all the effects.

Also, the locations as well as the timings seemed to be coordinated. Thus Wegener noted that the coasts of the Atlantic Ocean were "relatively free" from both earthquakes (seismicity) and volcanic action, whereas the coasts of the Pacific "are well provided with both."

There were other significant differences, too. The Atlantic coasts resemble "faults in a plateau," but the eastern Pacific coasts (off Asia) are bordered by island arcs and chains, to the east of which are found the impressive deep-ocean trenches.

Volcanic action rings the Pacific (see Figure 5-13). The most important thing about volcanoes, to Wegener, was the fact that they bring material to the surface from the dense rock layers of the mantle underlying the crusts. With this magma, or rock material, great amounts of heat also come to the surface.

Wegener, like every alert geophysicist, gave much attention to the temperatures under the surfaces of earth. He studied hundreds of temperature readings from both the continental and ocean floor crusts. On land, such readings must be taken at least 4 to 10 m under the top surface, in order to avoid the effects of sunshine, snow, or air temperatures above.

Volcanic action is so obviously energetic or even violent that it pushes certain questions to the forefront. If the temperatures underground are high enough to force to the surface millions of tons of molten magma or lava, plus debris, steam, and other hot gases—then do they also have sufficient power to propel continents, islands, and island groups across the face of the earth itself?

We have seen that Wegener willingly accepted the possibility that convection in the mantle rock substances might be the motive power for continental drift. In other words, the inside of the earth might somehow function as a great heat engine.

All about us on the surface of earth are heat engines, large and small, man-made to provide power for insatiable human wants in this technological period. They derive this power from such energy-rich fuels as gasoline, kerosene, oil, natural and synthetic gas, and even from coal and wood. But no such "burnings" take place within the earth itself. And there is no mechanism of pistons, turbines, gears, or shafts in the earth.

But is it necessary to have mechanical gadgets and linkages in order to convert temperature differences into mechanical motion and power? Not at all! The powerful winds that sweep the face of the earth do so as a result of temperature differences on the surface.

Imagine an old-fashioned stove sitting in a small room. So long as it remains unlit and at the same temperature as the air around it, nothing happens to that air. But if a fire inside makes that stove hot, air begins to circulate continuously. Around and around goes the air, forming a sort of single convection "cell" in the room.

What happens is simple. The air nearest the stove is raised in temperature, expands, and floats upward, its place being taken by cooler air drawn up from below. When the less dense air reaches near the ceiling, it spreads both ways and, having cooled, drifts downward again toward the floor. And so the circulation goes on, round and round, as a little smoke or dust in the room will make quite clear to an observer.

Convection is by no means a useless novelty. It makes possible very useful work. The engines of the famous old Model T Ford automobiles were cooled solely by convection, needing no water pump (Figure 8-1). Water in the jackets around the four cylinders grew hot, expanded, and became lighter, rose to the top of the radiator, then moved downward through it, cooling as it went. From the bottom of the radiator it was drawn back to the bottom of the water jacket around the cylinders, and so on.

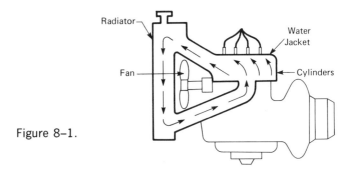

Figure 8-1.

Every time a pot of water or soup is placed over a small, high-temperature flame to heat, convection begins (Figure 8-2). It would be possible to turn a tiny paddle wheel by putting it into such a convection circulation.

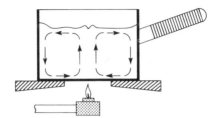

Figure 8–2.

Convection currents carry heat more rapidly from the high-temperature to the lower-temperature parts of such a system than the heat could travel by conduction alone—without the convective motion.

Early convection theorists—When Wegener wrote favorably of convection in the earth's mantle as a possible power source for continental drift, the notion was not entirely new. In 1919, R. Schwimmer had proposed "convection currents in the liquid layer [of the earth] caused by the non-uniform output of heat" within the earth. Such currents, Schwimmer suggested, would "draw the crust of earth along and compress it at areas where they [the currents] take a downward path."

Schwimmer's concept can be illustrated rather clearly, as in Figure 8-3. This pot of water is so large that two separate burners can be placed below it. The result is a double set of convection currents. At D2 in the center is a place where the currents "take a downward path." If we sprinkle small floating particles, such as sawdust, on the water, they will gather and clump together at D2. Some will also gather at D1 and D3, where convection currents descend at the sides of the pot. Thus the D locations show a compression, for the floating bits are pressed together by the "conveyor belt" action of the water currents under them.

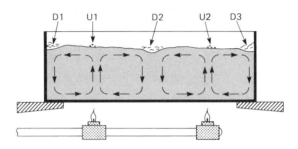

Figure 8-3. An extra-wide pot, placed over two separate flames, is shown here. Note that the relatively cool sites, where floating scum and froth collect, may be found elsewhere than at the sides of such utensils. Here, in the middle (marked D2), such accumulation also can take place. The D sites in this example are the relatively cooler collecting points for floating objects. The U sites are the hot spots, where the greatest amount of heat flow is found. It comes up from below with the moving liquid.

Also, two sites, labeled U1 and U2, mark where the upward currents affect the surface. At these places, tension or pulling-apart effects take place. Floating bits scattered in these D locations will be pulled away from them and carried to a U location by the conveyor belt action of heat-caused convection.

In 1925 another scientist, Otto Ampferer, declared that undercurrents of this kind had "dragged America westward," leaving behind the continents of Eurasia and Africa, in the great motion sometimes called the opening of the Atlantic Ocean.

But what on earth corresponds to the flamelike concentrations of high temperature that could cause heat convections under the crusts? The mythical idea that the center of the earth is a place of incessant fires, an inferno of the depths—this had long since been given up.

Wegener, however, studied a 1928 work by G. Kirsch called, in English, *Geology and Radioactivity.* Kirsch assumed that once all the continents were joined, but then radioactive atoms scattered under earth's crust had generated so much heat that thermal currents flowed below those crusts, "everywhere toward the ocean basins around them." The result: fragments of the supercontinent "were separated in all directions."

Leading authorities during the 1920s and later denied that dense subcrustal rock could develop the "relatively great fluidity" needed for such circulatory currents. Yet the concept appealed to Wegener. It was a way to account for the "split-up of Gondwanaland and also that of the single continental block composed of what is now North America, Europe, and Asia." Moreover, it would account for the "opening up of the Atlantic Ocean."

He doubted, however, that the state of earth sciences in the late 1920s permitted a decision as to whether or not the convection hypothesis had sound theoretical basis. The decades since then, however, have won an impressive number of leading geophysicists to a positive view—convection currents in the mantle or the outer parts of it are not only possible but strongly indicated by the accumulated evidence.

Tests of temperature—The "force" that starts and maintains heat convection currents is that of temperature differences. In much the same way, electric potential, measured in volts, is the force that starts and supports electric currents even through conductors with relatively high resistances. And pressure differences start and maintain the flow of gases or liquids in pneumatic or hydraulic systems.

In many situations, temperature differences exist with little or no noticeable flow of matter (convection). Yet even then another kind of flow does occur—the flow of heat energy through unmoving matter. It is marked by the gradual spreading of higher temperatures from the sites of highest temperature toward the lower-temperature regions. This kind of flow, free from detectable transport of matter, is called *conduction.* In many ways thermal or heat conduction is comparable to electrical conduction.

The electrical *conductivity* of a substance is its willingness to allow electrical energy to flow through it. The inverse or reciprocal of that conductivity is called *resistivity.* The willingness of a substance to allow heat energy to flow through it is called its thermal conductivity, symbolized usually by the italic letter k, and measurable in the modern unit-combination of power (watt) per length (meter) per degree of temperature difference. Such difference is measured either in degrees Celsius (°C) or in Kelvins (symbolized simply by K). The °C and the K are exactly equal in size.

Actual measurements of k show enormous differences in the thermal conductivities of different substances. This may be illustrated by an imaginary experiment (Figure 8-4). A rod with a uniform cross section of 10 cm^2 is mounted so that one end is heated by boiling water to 100° C, while the other is chilled to 0° C in a large block of ice. A thick jacket of cork insulating material covers the 1-meter length of the rod between the hot water tank and the ice container.

Sensitive thermometers show that a uniform temperature drop or gradient develops within the rod, from the 100° level at left to 0° at right. This gradient is 100° per m.

By measuring the average amount of ice melted per second by the right-hand end of the rod, we can roughly measure the amount of heat energy it conducts. To melt 1 gm of ice at 0° C requires 80 calories of heat, or 335.2 joule. Thus each gm per second of melting means 1/335.2 watt (W), or nearly 3 milliwatt (mW), of heat flow.

If the rod is made of silver, it should be found to transmit about 41 W of heat; if gold, 29 W; iron, 7 W; basalt rock, 0.22 W; granite, 0.19 W; glass, 0.1 W; and cement, only 0.03 W!

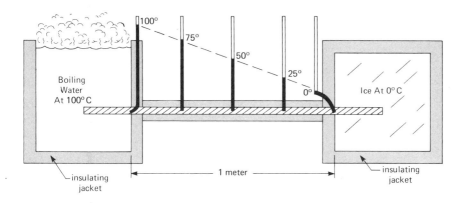

Figure 8-4. Measuring heat conductivity. In this schematic view, thermal conductivity is being tested in a rod of uniform cross section. At left it is heated to 100°C, at right cooled to 0°C. Separate thermometers along the intervening length show that an even temperature gradient, or slope, results. The insulating jackets around the rod prevent much heat loss to the surrounding air. The amount of ice melted at right reveals the amount of heat transmitted.

Some substances can convey heat power thousands of times as great as others can. The higher the transmission of heat flow, the greater the thermal conductivity of the substance.

These huge differences in heat conductivities reveal that metals are the best heat conductors, as they are the best conductors of electricity. On the other hand, some organic substances, such as paper, felt, plywood, and cork, conduct heat up to 10,000 times less "willingly" than do silver and copper.

The basic k figure for silver is about 410 W/mK, in units of the international system (SI). For basalt, 2.2; for granite, 1.9; for water, 0.6; for felt, 0.4; for cotton, 0.3; and for air, only 0.03. Typical thermal conductivities for substances common in the crusts of earth range between about 1.7 and 2.6 W per m per K or °C.

When we have measured this thermal conductivity symbolized by k and the temperature gradient, symbolized here by G, we can multiply the two together to find the resulting density of the flow of heat power—measured in units of the watt per square meter (W/m²).

Let us estimate this product (kG=W/m²) for the surface of the earth as a whole! Earth's average subsurface temperature gradient, G, is rather low or flat: about 0.03° C per m, or 3° per 100 m. The average k, or heat conductivity, of the substances in the relatively thin crusts of earth is about 2.1 W per m per °C. The product of this k and G comes to about 0.063 W/m².

That seems a very small rate of flow of heat energy. However, it equals 63 kilowatt per km², and since the earth's surface covers 510 million km², the total power of heat flow to the surface of our planet must be about 32 million megawatt. 1 megawatt (MW) is 1 million W or 1000 kW.

In using an estimated average figure for the geothermal gradients (G), we must not forget the important fact that the actual, observed temperature gradients vary, moderately but definitely, from place to place. In active volcanic and geothermal regions they may rise to 0.3 degrees per m. In most "normal" regions they tend to lie between 0.02 and 0.08 degrees per m of descent. Exceptional on earth are the "cold spots" where the temperature gradient is found to measure less than 0.02° C per m (2° C per 100 m).

Every substantial descent below earth's surface leads to levels with higher temperatures. But how uniform and consistent are these temperature gradients, or rates of temperature rise with distance descended? We surface-limited humans have barely nicked the crusts of the continents or oceans with the deepest man-made borings and drill holes. Yet geophysicists, using bold models and advanced mathematics, have tried to estimate temperature rises to the very bottom of the mantle (2,900 km down) or even to the heart of the core itself (6,370 km down).

The results here, as so often, upset what might seem to be common-sense assumptions. If the temperature gradients typical at, or near, the surface were maintained, the temperature at the center of the iron core should be more than 190,000° C! Even the bottom of the mantle, just outside the core, should be at about 86,000° C.

However, specialists on the thermal conditions inside the earth now agree generally that the central core temperature is no higher than 6,000° C, and many leaders in the field conclude that it is more like 3,000° C. Similarly, they calculate a lower mantle temperature no higher than 3,500 to 4,000° C, with strong support for a level even 100° or 200° below the 3,000° point.

In short, the temperature gradients in the near-surface crusts that our instruments can reach are enormously greater than those believed to prevail deeper down.

In fact, if we consider the entire earth from surface to center, the most probable average or overall temperature gradient appears to be much less than 1° C per km, compared with about 30° C per km, the general surface average for the earth!

Agreement is general that below about 100 to 200 km under the surface, the heat gradients flatten out and dwindle, continuing to dwindle all the way to the innermost core.

These are striking changes in the rates at which a descent, or a rise, of a given distance will be accompanied by a rise, or a descent, in the prevailing temperature of earth's substance. Such changes in the temperature gradients seemed significant to Alfred Wegener. He tried to interpret some of their principal causes and possible effects.

A leading cause, he concluded, was what he called "the crustal radium content." Modern scientists prefer to call it the crustal or near-surface concentration of radioactive substances in the earth. The particular element radium is actually only one of these substances which are now known to contribute to the earth's internal temperature patterns—and it is one of the least important in earth's total heating effects.

9. Radioactive Insights — and an Era of Eclipse

Understanding of radioactivity was quite incomplete when Wegener wrote and revised his book on continental drift. Just before the twentieth century began, the work of Henri Becquerel and of the Curies, Marie and Pierre, in Paris had revealed that strange and bafflingly energetic "rays" were constantly emitted by compounds of uranium, thorium, and of a new element that was named radium because of those same rays.

The British physicist Ernest Rutherford made historic measurements, revealing that these rays include high-speed subatomic particles and structures, as well as radiations related to light and to X-rays, but still more energetic. These outpourings of energy came from the hearts, or nuclei, of complex atoms. Incessantly, one by one, such nuclei fissioned, or broke apart, setting free the tiny bursts of energy. No such persistent outpourings of power had been found before from any substances or combinations of substances on earth.

The realization grew gradually that myriads of such radioactive atoms were scattered about to varying extents within the rocks forming the crusts and even the mantle of the earth. By the early 1920s it was becoming clear that these subterranean discharges of radioactivity might, indeed *must,* release heat within the earth in quantities sufficient to make important geologic differences.

Here Alfred Wegener's openmindedness was shown again. He

seriously considered such radioactive releases of heat as possible factors in the mechanism of continental drift. In his final edition, late in the 1920s, he clarified his own views of these possibilities.

If deep underground temperatures are far higher than temperatures just below the surface, then the interior of the earth clearly contains vast amounts of what is commonly called heat energy. However, Wegener believed that this could account for the great continental movements only if the upper layers of the earth's mantle were hot enough so they would flow or at least allow the less dense crustal plates to float through them. The temperature changes underground had to be such that the upper mantle layers behaved like a viscous liquid with respect to the relatively rigid continental crusts floating in and above them.

Drawing on the best available estimates of the era, Wegener offered his readers a multiple graph, like Figure 9-1. It shows five different possible patterns of temperature growth with increasing depth, the distances being below those men have even now been able to reach by their deepest wells and bore-holes.

For all five possibilities, Wegener assumed the same temperature rise within the first 20 km (12.4 mi) down: from the average surface temperature of about 14° C, an increase to about 500° at 20 km—that is, a rise of not quite 2.5° C per 100 m descent. This assumption agrees rather closely with the estimate of the British geophysicist E. C. Bullard that the temperature at the Moho, boundary between the crusts above and mantle beneath, is about 470° C higher than that at the surface.

But how does temperature change under the Moho, in the upper mantle? Wegener stated that this must depend on how the atoms of what he called "radium" were scattered within the mantle material.

By "radium" he meant all radioactive atoms. Modern geophysicists know that the most important ones for continental drift must be the following isotopes: uranium (U-238) with a half-life of 4.5

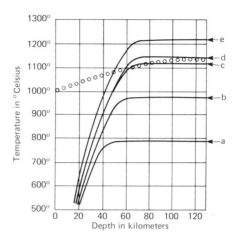

Figure 9–1. Possible patterns of increase in temperature (on vertical scale) with increase in depth (horizontal scale). Five curves, labeled (a) through (e), show the effects of varying locations of the bulk of the radioactive atoms that contribute much of the total heat constantly flowing from inside the earth. The dotted curve, which ends at right between temperature curves (c) and (d), indicates the lowest temperatures at which melting could be expected to take place.

Based on data from H. von Wolff, as used by Wegener in his book.

billion years; thorium (Th-232), half-life 14 billion years; and potassium (K-40), half-life about 1.4 billion years. The half-life is the interval within which half of any initial number of nuclei will have fissioned, or exploded, and so released their heat effects into surrounding matter.

Real radium releases heat far more powerfully, per gram or cubic centimeter, than any of these three isotopes, but the half-life of radium is only about 1,600 years. Hence within a mere million years a mass of radium will just about cease to emit appreciable radioactive heat. In round numbers, a mass of radium emits about one million times the heat power of like masses of U-238, Th-232, or K-40; but these are respectively about 3 million, 10 million, and 1 million times as long-lasting in their radioactive heat emission.

A mass of rock that contained one million U-238 atoms 4.5 billion years ago, about the time the earth itself was first formed or assembled, would still today contain 500,000 such atoms, not yet fissioned. Its rate of radioactive heat release would thus have dropped only by half in all those eons of time.

Wegener did not know all this but he was well aware that those radioactive substances that he lumped under the name of "radium" had marked chemical preferences for the granitic or sialic rock of the continental crusts, and less preference for the basaltic or simatic rock of the ocean floor crust. He did not know, however, the extent to which radioactive heat release was concentrated in each of the two crustal types, nor to what extent there might be still smaller rates of such heat release in the denser ultrabasic rock forming the mantle below the Moho boundary.

Even today geophysicists cannot provide exact answers as to the latter, but they can offer reasonably good estimates.

Wegener supplied five different possible curves for his graph, each corresponding to a different possible distribution of the radioactive atoms. Curve (a) assumed that they were concentrated rather close to the surface, down to about 45 km depth. Curve (e), on the other hand, supposed them to be more deeply scattered, down to about 65 km depth. The other curves—(b), (c), and (d)—represented possibilities between those of (a) and (e).

The calculations indicated that in the (a) case a temperature of about 800° C would exist 60 km down, whereas in the (e) case the temperature there would be much higher—about 1,150° C—and about 1,200° C at 80 km depth. In other words, the *less deeply* the radioactive atoms were bunched in the earth, the *higher* the prevailing temperatures should be at a given depth below the Moho boundary.

Why did Wegener stress these varied temperature possibilities? He noted that if either the (e) or (d) curve corresponded to the physical facts underground, then in a great zone from about 60 to 100 km down the mantle material should be so hot that it would melt, and "it is thus possible that here a molten layer is confined between two crystalline layers."

A molten layer would be one that flowed in response to forces sufficiently strong and long-lasting. Today, nearly half a century after these bold guesses of underground temperatures were made, it is widely agreed among geophysicists that there is indeed a deep layer, now called the *asthenosphere,* in the upper mantle, where matter is more *flowable* than either above or below. It lies very nearly where Wegener and his informants had supposed.

Why did Wegener believe that the earth's upper mantle, not far below the continental and oceanic crusts, must flow and yield to forces? First, it was the only way that he could conceive the continents could have moved such long distances. Also, evidence was strong that isostasy prevailed to a marked extent over the earth. This meant that the lighter crustal blocks forming the continents floated in the denser rock layers beneath, rising or falling until they attained equilibrium.

Hence the underlying mantle layers must permit vertical motions by the overlying crustal bits and pieces. And if vertical motions were possible through the mantle material, horizontal motions must be possible also. Thus Wegener pictured the great raftlike or ice-floe-like continents driven slowly through the denser mantle matter, much as a barge or scow is driven through the waters of a lake.

Gravitational shortages—Wegener called attention to some particular points on earth where, it seemed, isostasy did not truly prevail. These were sites where accurate measurements of gravitational acceleration showed strange deficits from the prevailing average.

Somehow, by a few parts per million, gravitational forces were weaker here than relatively short distances away.

Such was the case over certain unusually deep portions of the Pacific Ocean floor. Today these peculiarly deep zones are called *trenches,* as mentioned earlier in this book. We shall see that these trenches are assigned very special and surprising roles in the great cycles in which ocean crust is constantly being generated, shifted, and then swallowed up again within the earth's mantle.

Wegener mentioned the gravitational deficits over the Philippines trench, and the trenches near the islands of Yap and Guam and also near Hawaii. Curiously, as he noted, in each case there was a rise in the ocean floor not far west of the deep trench, and over the rise careful measurements showed an *excess,* rather than a deficit, of gravitational force.

Today, thanks to observations of orbiting man-made satellites, knowledge of the earth's gravitational field is far more exact than when Wegener worked. A satellite in circular orbit around the earth will reveal by moving faster that it responds to greater than average gravitational attraction toward the center of the earth, and by moving more slowly that it is passing over a region of less than average gravitational attraction.

Today, as will be shown in later pages, a great complex of deep ocean trenches has been charted, and a gravitational deficit is characteristic of each major trench. In the meantime, certain particular and peculiar locations have been mapped where the surface—ocean floor or continental crust, as the case may be—seems strangely raised. These uplifted or upward-bulging sites are likewise characterized by greater than average heat flow from the interior. In fact, nearly a score of such sites have been discovered around the globe and given the colorful name of "hot spots."

As Wegener considered the seeming violations of isostasy and the other oddities ("anomalies," in the language of earth sciences), he was impressed and hopeful, but remained cautious. In careful words he reminded his readers, near the close of his historic book, that "further research will be required before one can give a final verdict."

Even now, more than forty years after Wegener's death, a truly *final* verdict has not come in. However, preliminary verdicts are taking shape rapidly. The answers they suggest are of a sort that would have enormously excited Wegener and, no doubt, elated his faithful followers. To an extent that is truly impressive, they show Wegener had a fine "feel" for the physical facts that would be found fundamental for the branch of science now called global tectonics.

Yet it must be reported that Wegener's death in 1930 was followed by two decades during which he and his ideas received decreasing attention and were discredited or even derided at times.

The era of eclipse—Despite Wegener's patience and persistence, his continental drift hypotheses were not generally accepted, even during his lifetime. Support for his views dwindled markedly during the following quarter century. More and more established authorities and specialists in earth sciences turned away from his ideas.

Leading geologists at ranking universities and institutions of research regarded continental drift mostly as a swollen or even preposterous cluster of concepts. Geophysicists, by and large, regarded it as colorful but improbable. One of the most influential of the elder earth scientists, H. Jeffreys of Cambridge University, called Wegener's theories "insufficient" and "inapplicable"—and those negative comments were written in the late 1950s.

Wegener had exerted himself to draw on many scientific specialties for his evidence and arguments. Reigning specialists in those individual fields tended to disagree with him on relatively minor points in their own areas of competence, and then to assume that the remainder of his arguments also were faulty or misleading.

Some attacks on drift theory were almost savage. An especially emphatic foe was the Stanford University geologist Bailey Willis (1857–1949). The popular outline, *Fundamentals of Geology,* by V. Obruchev, a Soviet academician, says, "After first winning over many scientists, Wegener's hypothesis now arouses serious objections," and also, "Today's geologists regard the hypothesis ... as unable to explain the changes on the face of the earth."

Another prominent Soviet earth scientist, V. Belusov, listed factors which, in his view, refuted drift theory. He said it took no account of "oscillatory" or back-and-forth movements, though he regarded those as "important tectonic factors." He denied further that mountains had been formed by folding of granite continental crusts, for if the basaltic (ocean floor) layers below are softer than the granite layers, then it is the basalt rather than the granite that should fold. But if the basalt is harder—that is, more rigid—"then it should remain motionless."

Belusov granted that continents and ocean floors might differ in composition and structure, but he declared that the geological history of the earth showed them to be "tectonically identical." Finally, he complained of Wegener's inability to supply—that is, to describe —"the source of the forces moving the continents."

In a summary of the status of drift theory during the 1920–50 period, the eminent British physicist P.M.S. Blackett said that the theory had appeared to a few "as the obvious explanation of a wide range of phenomena; but to most it seemed wrong-headed and impossible." Meantime, the battle between the former and the latter "was often violent and sometimes almost abusive."

In *Debate About the Earth,* a useful book of the 1960s, three Japanese geophysicists—H. Takeuchi, S. Myeda, and H. Kanamori —wrote that drift theory had been "dismissed as Wegener's wild dream." They recalled that it "was hardly even alluded to in university lectures."

Truly, twilight seemed to have swallowed up the hypothesis of continental drift. It might have become merely an unimportant exhibit, a sort of freak, in the imaginary museum of discarded scientific conjectures—but for unexpected and dramatic developments during the years that followed the end of World War II in 1945.

A character in Shakespeare's comedy *As You Like It* speaks of "sermons in stones." Drift theory was recalled to life and growth largely because of new and almost incredibly sensitive methods for receiving and interpreting such "sermons." These methods were based on tiny variations in magnetic patterns that had been frozen into old stones of the earth, at the very time when they had cooled and congealed.

10. Magnetic Magic

The earth itself acts like an enormous magnet. That fact made possible the great navigational aid of the magnetic compass. But such a compass does not point to the true geographic north, except by accident. It indicates the "magnetic north," which is a long way from the North Pole. Where the author lives, a compass needle points about 15° east of true north. Near New York City it points nearly 10° west of it.

Also, one should not point toward the horizon when indicating the strict direction of magnetic north. The author, for instance, should point about 60° *below* the horizontal. A reader in northern Alaska should point about 80° below the horizontal.

The complex pattern of all such magnetic directions and intensities around the earth, or around any magnet, constitutes the complete "magnetic field."

Magnetic intensity anywhere is measured in terms of its direction, and in one of three units, very different in size. The *tesla* is the official unit of "magnetic flux density" in the International System of Units, but the *gauss* is still more widely used. 1 tesla (T) equals 10,000 gauss. And 1 gauss equals 100,000 *gamma*. Hence 1 T equals 1 billion gamma.

The earth's magnetic field is fairly feeble—as low as 0.25 gauss in parts of South America and rising only to 0.7 gauss at about 60° latitude in northern Canada. In contrast, a toy permanent magnet, shaped like a horseshoe, may maintain near itself fields of 20 or 30 gauss.

The earth is sometimes called a permanent magnet. But its field alters in intensity and direction. Seasonal changes alter the intensity by a few hundredths of 1 percent of the average anywhere on earth. There are also long-time drift effects. In about the year 1600 in London, magnetic north lay some 10° east of true north, by 1823 it had shifted about 35°—to 25° west of true north. Then it began drifting eastward, and by 1914 lay some 15° west of true north!

In about two thousand years the earth's magnetic field sweeps westward completely around the globe. Over far longer periods, changes far more drastic than this have taken place.

Soon after the end of World War II newly developed magnetometers proved themselves able to detect the magnetic patterns impressed on many rocks by the earth field that prevailed when those rocks were formed—that is, cooled. Rocks composed partly of iron oxides behave as though they contained myriads of tiny compass needles that freeze into position when the rock cools and congeals. Those molecular needles, aligned along the one-time field directions, no longer can shift within the rock. The ultrafaint field that they maintain around the rock reflects the long-past field of the earth itself.

Kinds of magnetic behavior—Some substances, called *diamagnetic,* are weakly repelled by an external magnetic field. Others, called *paramagnetic,* are attracted by outside magnetic fields. Some rocks, such as quartz, behave diamagnetically. Others, such as olivine or garnet, behave paramagnetically.

Some paramagnetic substances are so positive in their magnetic reactions that they acquire "spontaneous magnetization." They are called *ferromagnetic.* Iron is such a substance. Still another group are called *ferrimagnetic* materials. Many of these are nonmetallic and conduct electricity poorly, whereas the metals, including iron, are good conductors.

Both the ferromagnetic and the ferrimagnetic materials share one striking characteristic. If strongly magnetized by an external field, they retain some trace magnetization even after that outside field is removed. They seem able to recall their past magnetic experiences. Scientists call this memory *remanent* magnetism. It means about the same as "remnant" or even "remembered" magnetism.

Ferrimagnetism is typical of certain compounds of iron and oxygen that are prominent in many rocks, such as magnetite (Fe_3O_4). Such compounds are usual ingredients of the magmatic rock substances that well up from the mantle of earth, forming the sima rock crusts of the true ocean floors.

Such rocks acquire their remanent magnetic patterns as they cool. The directions of those patterns indicate the direction of the earth's field at that moment of congealing. These rocks thereafter reveal what scientists call *thermo*remanence.

From about the time of Wegener's death in 1930, to 1960, understanding of thermoremanence increased enormously. Measurements began to be made to determine the direction of the patterns impressed on such rocks found in various regions of the earth. Since the results revealed magnetic directions of the distant past when the rocks cooled, the study of these findings came to be called *paleomagnetism*.

Even some nonigneous rocks could be read by paleomagnetic methods, using the ultrasensitive magnetometers developed during the 1950s. Besides the igneous, magmatic rocks, there are also (as we have seen) the sedimentary and metamorphic rocks on earth.

The particles composing sedimentary rocks were commonly laid down under water. When these particles contained iron oxides, they had some tendency to align themselves with the earth's magnetic field as they swirled and tumbled downward. The effect is not likely to be more than 1 or 2 percent as strong as the remanence in igneous rocks. Yet it may be strong enough to be detected.

Metamorphic rocks were once either igneous or sedimentary rocks or both, before pressure, heat, or other strenuous conditions made them what they are now. Sometimes these transformations raised their internal temperatures so high that their iron oxide molecules were once again freed to align with the earth's magnetic field. Then as they cooled, they were locked into the new remanent pattern.

Thus it is true that rocks of each of the three principal types on earth may reveal former magnetic field directions to varying extents.

Probing paleomagnetism—Early in the 1950s a team of paleomagnetists from the University of London began studying the red sandstone strata scattered around much of England from the south to as far north as Newcastle. These sedimentary layers were laid down in the Triassic period, from about 200 million to 150 million years ago.

The results seemed utterly self-contradictory. About half the rock samples indicated a past magnetic north lying some 29° east of today's true north and about 34° *below* the horizontal. The other half indicated a past magnetic north some 39° west of today's true north and at an angle about 16° *above* horizontal.

The paleomagnetists—J. A. Clegg, M. Almond, and P. H. S. Stubbs—provided a simple answer to this paradox. The earth's magnetic field when the second group of stones were laid down had been the *reverse* of that in effect when the first group were laid down. During the 50 million years of the Triassic period the earth's magnetic field had reversed itself, apparently not just once but many times.

Even more remarkable for our purposes here, the magnetic lines of force, reversed or not, pointed differently by about 30°, or one-twelfth of a circle, during the Triassic than do today's lines of magnetism. Again the solution was simple but somewhat sensational: England itself must have rotated about 30° clockwise since Triassic times.

Similarly, the strange angles of dip shown by the remanent effects from those times could be accounted for if England had been much nearer the equator than it is now. Thus both rotation and long migration or drift were indicated by these sermons in stone.

Such strong hints of support for drift theory caused excitement in Japan and some other centers of geophysical studies. Paleomagnetists, chiefly British, continued, working on basaltic lavas that had cooled on the Deccan plateau of India between 150 and 30 million years ago. The actual ages of the rocks could be pretty closely estimated by radioactive dating methods.

Strangely, the older the rock, the farther south India seemed to have been positioned with respect to the magnetic field of the earth!

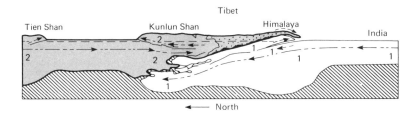

Figure 10-1. India as a great wedge, lifting up the mighty Himalayas and the highlands of Tibet. This view, used by Wegener in his book, shows the process as pictured by the French geologist Argand. The crustal plate of northern India, labeled 1, in moving northward, has been forced under the plate of southern Asia, labeled 2. The diagonally shaded zone below shows the underlying sima, or ocean floor crust. The white zones are the sial, or continental crusts, and the dotted sector shows sediments which were once part of the "Tethys Sea" that lay between Eurasia and the Indian subcontinent before the latter "collided with" and underran the former.

The downward direction of the Indian plate is clearly shown. Also apparent is the compression and folding of the Eurasia plate that produced the great elevation of the Kunlun Mountains.

Reproduced by permission of Dover Publications, Inc.

Rocks from about 150 million years ago seemed to have formed at about 40° *south* of the equator, those of 50 million years ago only a little south, and those of 30 million years ago somewhat *north* of that same equator. Apparently the great subcontinent of India had moved nearly 60° northward, or about one-sixth of the way around the globe. The average rate proved to be only a few meters per century, but the total distance moved was more than from Lima, Peru, to Montreal, Canada! Wegener, too, had insisted that India had moved a long way north before "bumping" Asia (Figure 10-1).

Strongholds of skepticism—No sudden conversions to continental drift followed these findings of the mid-1950s. How could one feel sure that these changes did not arise from large shifts of the earth's magnetic field with respect to its axis of rotation? It may have been the magnetic pattern that drifted, not the rocks.

A group of paleomagnetists associated with S. K. Runcorn (Figure 10-2) of the University of Newcastle upon Tyne, England, measured stratified formations in Britain and Europe, seeking the changes in the direction of magnetic north during the past 600 million years. They did find that the magnetic field had meandered widely during that time since the Precambrian era, when there was no life on earth. In fact, if one assumed that the Eurasian continent had not shifted much, then in Precambrian times magnetic north must have been located about where the tip of the peninsula of Lower (Baja) California is today.

The later wanderings of that magnetic north would have taken it southwest, to the central Pacific, by about 500 million years ago. Then it moved to the neighborhood of Japan about 300 million years ago, after which it crossed northwestern Siberia, and since completing that transit about 100 million years ago, moved to its present position.

Runcorn, in a 1955 paper, cautiously concluded, "Appreciable

Magnetic Magic 117

polar wandering seems indicated." But, even more carefully: "There does not yet seem to be a need to invoke appreciable amounts of continental drift to explain the paleomagnetic results so far obtained."

Even India's seeming northward drift of some 7,000 km could be disposed of if the earth's magnetic north and magnetic field had shifted southward to a corresponding extent.

Runcorn and his colleagues then investigated further in North America, including sedimentary rocks in Texas and Arizona. Within two years they published their picture of the migrations of the earth's magnetic north as "seen" by rocks formed in the New World. Then

Figure 10-2. S. K. Runcorn, eminent geophysicist. He and his associates made historic contributions to the paleomagnetic studies that revived continental drift theory and helped to extend it to a far larger scope, in the form of the modern concepts of global tectonics.

they compared this picture with that obtained from the rocks they had measured in Europe and the British Isles.

They found rough correspondence, but a good bit *too* rough. More detailed analysis revealed that between about 600 and 200 million years ago the two separate tracks deviated to the extent of some 30° in the direction for magnetic north. Such a discrepancy was as great, for example, as the distance from the northernmost islands of Japan to Unalaska in the Aleutians.

Besides, this discrepancy seemed to fade out since some 60 million years ago. Rocks formed since then showed harmonizing magnetic directions, whether formed in Europe or North America. A simple but insistent question remained: What possible event could account for a 30° difference of latitude lasting 400 million years, then dwindling to nothing between 200 million and 60 million years ago? If North America had moved—drifted—away from Europe to the extent of 30° of longitude during that 140-million-year interval, then the two separate tracks or curves could be made to coincide. Not otherwise.

This time the evidence of the rocks could be explained neither by polar wanderings alone nor by continental drift alone. Only a combination of both could harmonize the remanent records!

Continental complications—Rocks on other continents have been similarly studied with precision and patience. As each revealed its own picture of the apparent wanderings of the magnetic poles, the record became even more complicated. In general, in order to reconcile all the remanent findings, it became necessary to visualize the continents as rotating to various extents as well as drifting.

Evidence for a 30° turning of the British Isles has been mentioned. The islands of Japan appear also to have twisted clockwise by an appreciable amount. Moreover the rock records suggest that about 60 million years ago Japan was bent in the middle, for its present southwestern half seems to have turned clockwise about 40° more than its northeastern half.

Paleomagnetic methods, used alone, have difficulties and limitations. Yet by the beginning of the 1960s paleomagnetism had placed continental drift ideas squarely in the spotlight of all the major earth sciences. Magnetic magic had ended the era of eclipse for the great Wegenerian hypotheses!

11. Creating, Carrying, and Consuming Crusts Under the Oceans

During the 1950s and 1960s the theory of drift was supported and enlarged by evidence from additional directions. Ocean bottoms were measured and mapped more precisely than ever before by means of automatic sounding devices. Tiny variations in gravitational force were detected over ocean floors and continental sites. Needlelike temperature sensors were thrust into the ocean bottom and into land surfaces in selected locations.

Strange as it may seem, even in the late 1950s most of the striking features of the floors of the world's oceans were hardly known or even suspected. Bit by bit the amazing submarine picture was assembled, abounding in surprises and paradoxes.

Oceanographers knew that, for some unknown reason, the mid-Atlantic about halfway between its eastern (European) and western (American) shores became markedly shallower than elsewhere. They knew, too, of various islands of volcanic origin scattered about the mid-Atlantic, from far south to far north of the equator. Among these islands were Bouvet, more than 50° S latitude; Gough, more than 40° S; Tristan da Cunha, between 30 and 40° S; and St. Helena, more than 15° S. About 40° N latitude are the Azores. And just below the Arctic Circle lies Iceland, rich in volcanoes and outpourings of heat from the earth.

Surveys of the ocean floors have revealed that through or near such islands runs the long and massive mid-ocean ridge, often called the Mid-Atlantic Ridge or Rise (Figure 11-1). It closely duplicates the curves of the coasts far east and west of it. Francis Bacon doubtless would have been enchanted by another "conformable instance," could he have been shown these tell-tale contours deep under the sea!

Alone vast stretches of the rise, the summit is slotted by a great rift valley, gouged out like a groove. Here the heat flow is far higher than normal for the ocean floor as a whole. In places it rises to as much as six or eight times the normal rate.

It is quite as if this rift in the mountain chain were atop a crack in the ocean floor crust, through which heat seeps from the hot mantle substance below. The rift valleys are extraordinary also in another way. A large number of earthquakes are found to originate under them—and not very far down, as earthquake origins go. The subrift quakes, in particular, are less deeply sited than the earthquakes, usually under continental margins, around the great "ring of fire" that borders so much of the mighty Pacific Ocean. Those deeper quakes are found to lie beyond—usually some distance west of—the deep ocean trenches that lie alongside the Pacific coast of Asia.

The mid-ocean rises clearly have some important geophysical function or action, resulting in the unusual heat flow, the shallow earthquakes, and the impressive extent of the rises themselves. In fact, taken as a whole, they form the earth's greatest chain of mountains—mountains invisible only because their summits lie far below the surface of the oceans.

Besides the great Mid-Atlantic Rise, running from Iceland to 40° or more S latitude, there are related ridge systems curving eastward around the southern tip of Africa, then running northward all the way through the Indian Ocean into the torrid Gulf of Aden, nearly 15° N of the equator.

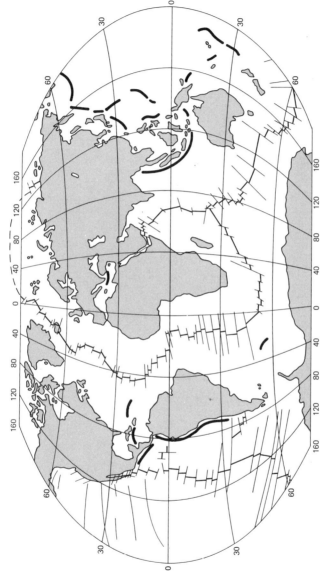

Figure 11-1. World's principal ocean rises show here as medium-weight lines. Finer lines crossing them are great fault-fractures, offsetting the rises. The East Pacific Rise runs from west of South America all the way into the Gulf of Lower California. Heavy lines show oceanic trenches. Most of these border the Pacific. Others appear in the far south Atlantic, also east of the West Indies, and even in the Mediterranean.
From D. H. and M. P. Tarling, *Continental Drift* (London: H. G. Bell & Son, 1971; U.S. edition by Doubleday).

Creating, Carrying, and Consuming Crusts Under the Oceans 123

This truly worldwide network of oceanic rises has been said to loop around the earth like the seams on a baseball. Its combined length is about 40,000 km (25,000 mi), roughly long enough to encircle the entire earth at the equator. Its submarine summits soar kilometers higher than the ocean floor level some distance in either direction from the center line of the rift or summits. In sheer bulk as well as length these submarine mountain chains are imposing. But in other ways they are even more extraordinary.

Precise soundings by means of waterborne sound waves have shown that the so-called rift valley, where it runs along the summit of the Mid-Atlantic Rise, is about 2 km deep and up to 50 km in width (Figure 11-2). It directly overlies the belt where the earthquakes under the rise are concentrated.

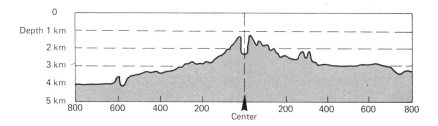

Figure 11-2. A typical cross section of the Mid-Atlantic Rise is suggested here, with the vertical scale multiplied to eighty times the horizontal scale. Sea level is at the top, marked 0 km.

The central rift is marked by the broken line.

On either side of this strange central split lie the peaks of the main range, and further out along the slopes are the lesser peaks.

The floor of the central rift lies far lower than the main peaks. In fact, the sloping sides of the rise do not get down to the level of the rift floor until more than 150 km from that rift. Clearly, powerful forces must be at work to pile up these enormous rise systems and split them by these vast rifts.

Based on data published in 1957 by C. H. Elmendorf and R. C. Heezen.

Similar rifted ridges run along the floor of the Indian Ocean and midway between the continents of Australia and Antarctica. Another ridge curves along the floor of the eastern Pacific Ocean, in a generally northward direction. Each of these mid-ocean ridges runs very nearly halfway between the limits of the ocean on whose bottom it is found.

The ridge-top rifts are regarded as gaps or pulling-apart sites, because the rises as a whole lie on top of the ascending currents of great subcrustal convections. Hence, the "conveyor belt" effect creates tension—the plate on one side of the rift is constantly being pulled away from that on the other side of the rift (Figure 11-3).

However, the ridges and their central rifts are repeatedly offset by great faults or fracture lines, running very nearly at right angles to the ridge line at that point. Such great ridge fractures may be traced as far as 100 km or more from the ridge summits.

The sophisticated echo-sounding systems that now map ocean bottoms are able to distinguish the approximate depths of layers of sediment that cover the stone crust under the ocean. For about 100 km in either direction from the central rift, the slopes of the rises are almost bare of sediments. Beyond that distance they slant down under the sediments, which become deeper with increasing outward distance from the ridge. Since subocean sediments accumulate with time, it appears that the crust closest to the central ridge-rift is the newest or youngest, and the crust becomes older with greater distance away from the ridge. In fact, when samples are gouged out of the basaltic rocks of the ridge systems they are typically found to be as young as 10 million years—almost newly born, in geological terms!

Using radioactive dating of crustal rocks and of the sedimentary layers above them, the general principle has been established that the ocean floor becomes older with increasing distance from the ridge

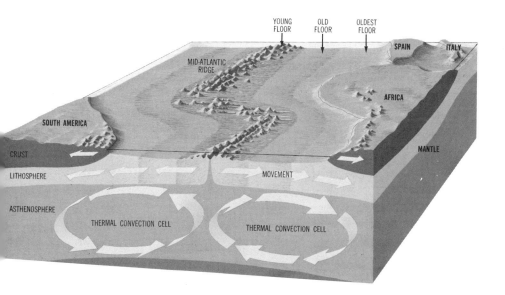

Figure 11-3. Ocean-floor spreading in action. Newly formed crust moves outward from the mid-ocean rise or ridge, in this case the great ridge midway between Africa (right) and South America (left). The process is the result, not the cause, of the great convective cycles taking place below in the asthenosphere, the mobile upper part of the earth's mantle.

The result—young ocean floor lies closest to the ridge, older floor less close, and the oldest floor most distant. In the Atlantic the oldest ocean floor is found closest to the continents on either side of that ocean.

Note that at right the continental crust of Africa moves with and above the ocean floor crust below it, and so, too, does the continental crust of South America with the ocean floor crust below it.

Illustration by Jim Schick, from *ESSA,* April, 1970, magazine of the Environmental Science Services Administration, U.S. Department of Commerce.

summit or rift line. This relationship is possible only if new ocean floor is being produced in the ridge-rifts and then being moved out in either direction.

Also, remanent magnetism in the basaltic crusts of the ocean reveals patterns by means of which the rock age can be estimated in a manner undreamed of when Wegener worked.

A ship on the surface of the ocean, or even a plane flying above that ocean, tows a sensitive magnetometer, at the end of a long line in order to keep it away from disturbing magnetic metals in the towing vehicle. As it moves, the magnetometer responds to the general magnetic field of the earth and also to the remanent magnetic pattern of the basaltic rock in the ocean crust below. Tiny but unmistakable jumps in magnetic intensity register from time to time. A downward jump indicates that the remanent magnetism of the ocean floor below, which had been reinforcing the general magnetic field of the earth, has suddenly reversed, and now is opposing or bucking that field. A later upward jump means that the subocean crust is again magnetized in a manner that reinforces the prevailing field of the earth.

These reversals, taking place repeatedly as the plane or ship moves outward from the ridge summits, show the reversals of the earth's magnetic field that took place in bygone ages, when the rock below cooled and its remanent magnetic pattern became fixed.

If the ocean floor is mapped so that "normally" magnetized zones are colored black and the reverse-magnetized zones left white, the floor shows enormous stripes—and the pattern of black and white from the ridge line eastward (for example) is duplicated in reverse order from that ridge line westward. In other words, when reverse-magnetized rock moved out in one direction, that same kind of rock moved also in the opposite direction. And when "normally" magnetized rock went one way, it also went the other way.

The process, in fact, is like that of two enormous magnetic tapes emerging from a vast recorder, each impressed with the same magnetic pattern, but one going one way, the other the opposite way.

The radioactive dating of rocks and sediments has enabled paleomagnetists to construct a calendar of past reversals of earth's magnetic field. The present magnetic field direction is understood to have prevailed since some 700 thousand years ago. But repeatedly before then there were flip-flops or reversals of the magnetic field—in fact, about thirty-two during the past 80 million years, or an average of one every 2.5 million years.

These reversals are sometimes called "events," sometimes magnetic "anomalies." Their cause remains largely a mystery, but there is no doubt as to their reality. The earth's magnetic field is widely believed to result from the flowing of conducting metal—liquid iron —within the earth's core. Turbulence in the pattern of that flow may cause the reversals, in which what had been the magnetic north becomes the magnetic south, and vice versa.

The timing of these magnetic flip-flops has been coordinated with the distances from the central ridge to the part of the ocean floor where the magnetic reversal is recorded. The result makes possible estimates of the rate at which the newly formed ocean floor crust has emerged from the rift and spread out.

For example, suppose that a boat or plane tows a magnetometer westward and at a distance of 10.5 km encounters the magnetic drop linked with the reversal of 700,000 years ago. This shows that the average speed of crustal emergence has been 1.5 cm per year—if the word "speed" can be used for a rate so very slow.

In the opposite direction also, corresponding reversal of remanent magnetism is found 10.5 km from the ridge. Hence the total rate of new ocean floor emergence from that part of the ridge has been 1.5 plus 1.5 or 3.0 cm per year. In other words, 3 m per century, 30 m per millennium, or 30 km per million years.

Above the great fracture zones, such magnetic measurements are hard to make, for the offsetting confuses the magnetic effects. However, working between the fractures, with ultrasensitive magnetometers and speedy towing vehicles, geophysicists have accumulated enormous amounts of truly astounding data.

Here are approximate ocean floor spreading rates from seven great ridge systems, in units of meters per century. (a) North Atlantic, 2 m; (b) North Indian, 2.5 m; (c) South Atlantic, 3 m; (d) South Indian, 6 m; (e) North Pacific, 6 m; (f) Pacific-Antarctic ridge, 8 m; (g) East Pacific, 16 m.

Thus the East Pacific Ocean floor is spreading about eight times as rapidly as the North Atlantic floor. The South Atlantic itself is widening about 50 percent faster than the North Atlantic.

The extremely active East Pacific Ridge can be traced approximately on a globe of the earth, from Easter Island (belonging to Chile) through Clipperton (French possession) about 10° N of the equator, and then all the way into the Gulf of California. Its rapid creation of new ocean floor crust can be linked with the heavy earthquakes along the west coast of South America, some distance east of the ridge itself.

How can we visualize such an oceanic ridge in cross section? Figure 11-3 suggests a simplified view. We see that convection currents below the crusts pull them apart by a "conveyor belt" action. Into the resulting opening the hot magmatic material of the mantle flows from below.

How much new ocean floor?—Fred J. Vine, a British geophysicist, has estimated that half of the present ocean floor crust was created during the past 65 million years. This is less than 1½ percent of the earth's probable age!

During a million-year period, the total of new ocean floor that emerges from the ridge systems of the world must be between 2.5 and 5 million square kilometers, amounting to at least 1 to 2 percent of the present total area of true ocean floor. Meanwhile, the area of the continents is not becoming less. In fact, there is good reason to believe that the continental areas are increasing slowly but steadily.

Hence, since the radius and total surface area of the earth almost certainly are not becoming larger, there must be some other process at work somewhere for eliminating the older ocean floor crust! Nothing, neither crusts nor sediments, from the ocean bottoms is found to be *old,* geologically speaking. The most ancient sediments or rocks from ocean bottoms are youthful, almost juvenile, compared with typical rocks widely found on continents. What is the secret of the ocean's seemingly eternal youth?

Other extraordinary evidence gathered above the world's ridges has been studied in the effort to find the answers. Directly above the ridges, the force of gravitation shows unmistakable increase. It is as if the welling up of hot matter from below, which makes the heat flow exceptionally high, also makes for greater masses of matter. Elsewhere, we have seen, isostasy operates to equalize the effect of mountains or valleys to a large extent. But under the ridges something is checking or counteracting isostasy.

A look into the trenches—Earth's oceans also contain other strangely opposite features, in many ways the reversals of the great ridges. They are usually called the deep-ocean trenches or troughs. Besides being exceptionally far below the surface of the ocean, they show lower than normal heat flow from their submarine surfaces. Also, over the deep trenches, gravitational force is unmistakably less than normal.

The strange locations of these trenches are suggested in Figure 11-1. They form a spectacular and somewhat irregular broken ring around the Pacific Ocean. Indeed, their placement associates them with the Pacific's "ring of fire," that vast pyrotechnic zone of earthquake activity and volcanic action which rims the planet's largest single geographic feature.

The system of deep-ocean trenches can be traced starting at the southern end of the west coast of South America. From there the huge Peru-Chile Trench runs north to Colombia. Then the Middle American Trench begins near northern Panama in Central America, and runs northwest along the coast to about the latitude of Mexico City.

A great trenchless gap follows. No true deep-ocean trench appears all the way up the Pacific coast until south of the Alaskan peninsula and the Aleutian Islands, which arc down to 50° N latitude, like a great string of beads swung from Alaska toward eastern Siberia. All along the southern side of this mighty curve lies the great Aleutian Trench, reaching depths between 6 and 9 thousand m (20 to 25 thousand ft) below sea level. Near the southern end of Kamchatka Peninsula, the trench system veers sharply southward. The new direction is traced by the Kuril Trench, which attains depths of more than 10 thousand m before it terminates just east of the gap between the northern Japanese island, Hokkaido, and the southern one, Honshu.

The Kuril is followed by the Japan, the Izu, and the Mariana trenches. The latter curves east of the Mariana Islands and attains the greatest oceanic depths on earth—fully 11 thousand m (more than 36 thousand ft) below sea level!

From here the trench system moves southward, double rather than single. East of the Mariana is the Ryuku Trench, ending near Taiwan. East of the Philippines lies the Philippine Trench, pointing toward New Guinea. Pointing also in that same general direction, but further to the east, is the relatively short Yap Trench.

Then, swinging toward the southeast, the New Britain Trench runs between New Guinea and the Bismarck Archipelago. East of its southern end is the Vityaz Trench, and south of that same end, the New Hebrides Trench, whose southern end extends toward the Fiji Islands.

From west of Samoa to the northeastern point of New Zealand can be traced two linked trenches: the Tonga and the Kermadec.

Scope and capacity of the trenches—This ocean-surrounding trench system totals tens of thousands of kilometers in length. Its significance is unmistakable in modern global tectonic theory. Just as the ridge-rift systems produce and emit newly born ocean floor crust, so do the trenches swallow up and eliminate old ocean floor crust moved toward them by the earth's great heat-powered processes. On the transmission belt of convection the ocean floor crust is forced toward, under, and finally downward away from the crust of the great tectonic plate that rides above (Figure 11-4).

Geophysicists have invented a word for this process—*subduction,* signifying "drawing down." The crust of the Pacific plate is subduced under the Eurasian plate in the trenches that line the eastern margins of Asia. The crust of the East Pacific plate is subduced under the trenches that line the western margins of South America, and so on. The rate of such ocean floor consumption is considerable by geological standards. It is believed to attain 8 or 9 cm per year in the Kuril, Japan, and Mariana trenches. That is equivalent to between 5 and 6 mi per million years.

In the great Tonga Trench east of Fiji, Pacific Ocean floor crust is consumed, and this takes place beneath the tectonic plate that carries New Zealand and Australia. The Tonga Trench, too, seems to devour ocean floor crust at about 8 or 9 cm yearly. However, the southern Kermadec Trench seems to dispose of ocean floor crust at only about half that rate.

Far to the north, the Aleutian Trench appears to consume Pacific crust at between 5.3 and 6.3 cm per year, while the Java Trench, southwest of Java and Sumatra, swallows up the Eurasian plate under the Australia plate at a rate of about 4.9 cm per year.

Some authorities have concluded that the total consumption of

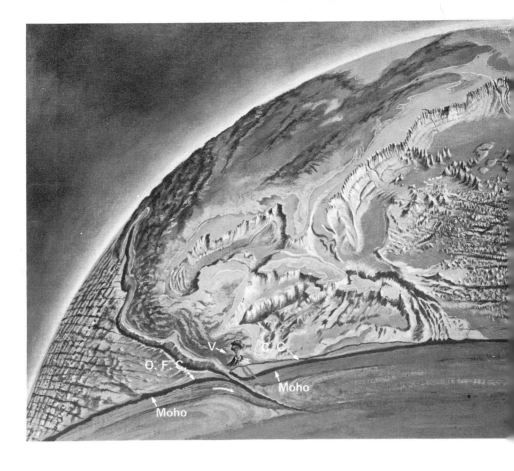

Figure 11-4. Subduction at work below a trench. The ocean floor crust (O.F.C.) is pressed against the continental crust (C.C.) and forced to bend below it. The crusts here are the relatively thin dark layers riding atop the upper mantle lithosphere and moved by convection currents in the more mobile asthenosphere below.

On the surface near the subduction area can be seen Nicaragua, El Salvador, Honduras, and so on. Curving along the west coast of Central and North America is a dark line—the bottom of the trench as seen from above. This bottom lies several kilometers below sea level now.

The downward slant of the subduced oceanic crust is the site of earthquakes, some of which take place as deep as 600 or 700 km below the

Creating, Carrying, and Consuming Crusts Under the Oceans

surface. V indicates a volcano, formed by the hot magma and vapors emerging from the mantle far below.

The descent of old ocean crust toward the depths is thus accompanied by spectacular seismic, volcanic, and thermal effects. New ocean crust is formed from upflowing hot magma along mid-ocean rises of the world, and the return of such crusts to the depths causes frictions and fissures which also remind us that our earth is hot below.

Adapted from an illustration in *Horizons,* magazine of the Amoco Production Co., Tulsa, Okla.

Pacific Ocean floor crust may be greater than its production by the ridge systems lying west of South America. The net result may be that the vast Pacific is gradually narrowing. Some day, in fact, the Atlantic, which did not even exist as an ocean some 200 million years ago, may be wider than what is left of the enormous Pacific!

12. Total Tectonics

Can any great generalizations be drawn from all this—from ocean floors that emerge from submarine rises, spread far across the face of the earth, and then plunge below continental crusts, back into the consuming mantle within the earth? About 1967 such vast generalizations began to be proposed. Together they form a system of understandings or insights that is called by such names as "global tectonics" and "plate tectonics."

Here we call it "total tectonics," for it gives new meaning to major physical events all around and within the earth now and during long ages past.

The word *tectonics* means the art or technique of assembling, shaping, and otherwise preparing materials for use in construction—architectural construction, particularly. In geology it has come to have a special meaning, referring to the conditions that cause motions of the earth's crusts, and the structural results of such crustal movements.

Able and versatile young geophysicists took the lead in sketching and revising the new tectonics. They broke boldly away from many past assumptions. These young rebels included such "tectonicians" as W. J. Morgan (Princeton University), Xavier Le Pichon (Lamont Geological Observatory), R. L. Parker and D. P. McKenzie (Cambridge University), and many others deserving of mention, but for limitations in space here. The remaining pages of this book seek to present highlights of their ideas, which together form a truly revolutionary way of approaching the earth sciences.

The entire earth is jacketed by a number of relatively rigid tectonic plates, sometimes called blocks. Each of them moves as a whole. The total number of such separate plates is now at least twelve and may be as high as twenty. The more detailed and accurate the data, the more exactly it will be possible to trace the maximum number and the outlines of the separate plates.

These plates are slowly shifted by currents of a convective origin, flowing in the asthenosphere, the mobile layer in the earth's upper mantle. The convections are the direct result of the heat released into the interior of earth by radioactivity, which results in temperature differences leading not only to heat flow by conduction but also by actual movement of heat-expanded matter—known as convection.

The tectonic plates encounter each other at their boundaries. It is there that they impinge on one another and interact. And it is at the great plate margins that we find the geophysical events which matter most in determining the changing face of the earth.

Three major kinds of plate encounters are possible. First is the separation or pulling apart of adjacent plates, resulting typically in mid-ocean rises and the oozing upward of mantle material to form new ocean floor crust of basaltic (simatic) character. This is the situation that brings about creation and spreading of new ocean floors.

The opposite of this pulling apart is the pushing together or compression of a pair of plates, head-on, so to speak. This typically brings about the overriding of one plate—commonly the continental plate, which is less dense—and the underthrusting of the other, usually the ocean floor plate, which is denser. That is the origin of the deep sea trenches, into which ocean floor crust is continually thrust while still relatively young. It bends over, following a downward convection path, and slowly passes into the mantle where, so far as is now known, the heat softens and finally melts it, so that it becomes indistinguishable from the remainder of the mantle substance.

A third great possible plate-to-plate encounter is the lateral motion, the sideways sliding of one plate boundary along the other. We see this along the two sides of the San Andreas fault systems that run a little inland from the Pacific coast in the far western U.S.A.

It has been said that the three great plate-to-plate encounters duplicate the three great aspects of deity in the Hindu religion: Brahma, the Creator, corresponds to the pulling apart of plates with new crustal emergence; Siva, the Destroyer, to the thrusting together of two plates, with resulting destruction and consumption; and Vishnu, the Preserver or Maintainer, to the lateral shifting of plates, neither of which is increased or diminished in the process (though the attendant earthquakes may seem destructive enough to the tiny humans who inhabit the shifting plates!).

Continents and oceans both ride the plates—Basic to total tectonics is the understanding that the typical great plate carries both continental crust and ocean floor crust. Only the Pacific plate and one or two minor plates are composed entirely of a single type of surface—ocean floor crust.

The same great plate that carries North America also carries the western half of the North Atlantic Ocean. On the other side of the Mid-Atlantic Rise, the same plate that carries the eastern half of the ocean floor carries also Europe and Asia. (China may lie on a plate separate from the remainder of Asia, but that is a detail needless to explore here.)

The South Atlantic shows the same carrying of both ocean floor and continental crusts on a single great plate. The South American plate carries also the ocean as far as the Mid-Atlantic Rise, while east of that rise the ocean floor rides on another plate which carries the continent of Africa.

All this is the consequence of the way the continental and ocean crusts are formed and the ways that the plates are moved. In the first place, the continental masses are the slag or froth of the smelting processes carried on within the earth. The convection currents that carried the continental granites to the top also massed them together. Later convective forces split apart and separated continental masses, as when Africa and South America were torn asunder. The growing separation became an intervening ocean, whose new floor emerged continually from the rise system that remained roughly midway between the opposite shores of the new ocean.

Hence, an "old" continent may be bordered by a "new" ocean floor. But the juncture between the plates will not be where the ocean floor crust meets the continent. Rather, it will be midway within the ocean floor crust.

Quite different is the head-on collision situation. Here the ocean floor crust typically follows the downward convection, passing under the lighter continental crust. The continents ride higher and survive; the ocean floors return to the depths while still geologically new or young.

The slow splitting apart of great continental masses is very likely going on even now. Strong evidence indicates that the eastern part of Africa is being stretched and separated, at and southwest of the Red Sea. Great rift valleys may be traced there. Future ages may study how once the separated continental segments were one. The Red Sea may be a new ocean in embryo.

Comparable splitting is pulling apart Lower California to the west from the Mexican mainland to the east. It is like some great fun-making machine in an amusement park: everything turns, twists, churns, and changes.

Continuing continents—Amidst the long cycles of shift and change, the continental masses on the whole seem to enjoy a charmed life. Continents may split and separate into parts. Parts may join, as when the Indian subcontinent traveled a long way north and became part of the rest of Asia. However, the total continental area does not dwindle. Indeed it probably increases slowly, for continental crust floats safely on top, much as Noah's Ark floated safely above the waters of the great deluge.

As oceanic crusts are thrust below continental crusts, at the deep ocean trenches, it is likely that great masses of sediment are scraped off the descending ocean floors. These masses of sediment may in time be added to the nearby continent, thus increasing continental area still further.

The principal processes of total tectonics are shown in a single imaginary view here (Figure 12-1). Below R we see three segments of a ridge system, each separated by a great fracture or fault line from its neighbor. From each segment new crust emerges and spreads out in both directions from the rise. The crust is labeled *lithosphere,* to show that it is part of the realm of rigid rock lying above the *asthenosphere.* The latter is the mobile layer of the mantle, and its motions determine the motions of the lithosphere above it and of the crusts above the lithosphere.

Wegener imagined the crustal rafts as being pulled or forced through a stationary liquid or semiliquid layer. Modern tectonic theorists, however, visualize great currents, convective in nature, flowing within the asthenosphere and bearing along with them, like ice floes or great rafts, the overlying lithosphere and the crusts above it.

At the right of the diagram, below T-3, ocean floor crust and the lithosphere below it are propelled downward into the mantle at an oceanic trench. Above it remain, at far right, the lithosphere and the crust under the continent. This situation is repeated along thousands of miles of trenches east of Asia and west of South America. It is subduction at work.

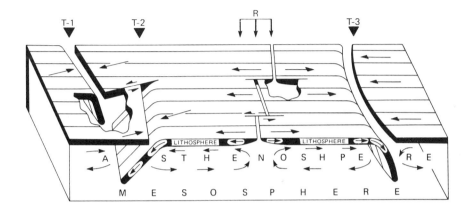

Figure 12-1. Imaginary view of typical plate motions as seen by the new "total" tectonics, interpreting the movements of the principal features of the earth's surface. Different kinds of compressional or coming-together movements are shown below at T1, T2, and T3; while below R, tensional or moving-apart motions produce spreading actions.

The crusts of the earth, formed into separate plates, make up the relatively cool lithosphere on top. Under it lies the weaker asthenosphere, or upper mantle, in which temperatures from 500 to 1000°C make possible the "creep" or flow powered by temperature differences and the resulting alterations in rock expansion and densities.

Adapted from B. Isacks, J. Oliver, and L. B. Sykes, "Seismology and the New Global Tectonics," *Journal of Geophysical Research* (September, 1968).

At the left, a double subduction is shown. T-2 indicates the left-moving plate being subduced below the other. T-1 indicates the opposite situation. From the near end of the latter to the far end of the former trench runs a great fault or fracture line. This is the kind that geologists call an "arc to arc transform" fault. Other varieties of fault, not pictured here, are called "ridge to arc" and "ridge to trench" faults.

Tectonicians remind us that a plate always moves away from a ridge (rise) and toward a trench, never the other way around. The ridge, indeed, is formed by that "moving away"; the trench by the results of that "moving toward." Finally, along great faults the rubbing of plate against plate produces the jerky or spasmodic motions that we humans call "earthquakes."

Tremendous turnings—So far we have considered plate motions as if one plate moved always in a straight line relative to another. Actually, this is oversimplified and in fact misleading. The plates that jacket the earth are all segments of the surface of a great sphere. Relative movements of two separate segments on a sphere are bound to be *rotations* of some kind. Seen from one plate it is as if the other were hinged to an axis projecting through the sphere. It need not be the same axis of rotation that we visualize as running from the earth's North Pole to its South Pole.

The mutual rotations of segments of a sphere can be described by locating the poles of the axis around which they seem to turn, and determining the rate—as in degrees of angle per second or year—of the turning motion.

Late in 1968 a useful study of such relative rotations of the tectonic plates of the earth was published by Xavier Le Pichon (Figure 12-2), already mentioned. He limited his analysis to six principal plates, disregarding, for the time being, various subplates which later and more detailed data will doubtless permit to be included.

The relative rates of rotation of the six major plates were found to be quite small—ranging from about 2.5 to 12.5 degrees of angle per 10 *million* years. Yet so large is the sphere of the earth and so extensive are the plates that these tiny rates of turn produced mo-

Figure 12-2. Dr. Xavier Le Pichon, a pioneer analyst of the relative motions and rotations of the great tectonic plates that form the surface of our earth.

tions of separation of up to about 12 cm (4.7 in) per year along plate boundaries, and motions of approach or compression of nearly 11 cm (4.3 in) per year.

A world map showing the six plates appears here (Figure 12-3). We can identify each by a capital letter—A (American plate), E (Eurasian), F (African), I (Indian), P (Pacific), and T (Antarctic). These six plates encounter each other in twelve different situations, of which five result in separations or pulling apart, and seven result in approaches or pushing together.

The A plate here includes North, Central, and South America, as well as the Atlantic Ocean floor west of the Mid-Atlantic Rise. The E plate includes Europe and almost all Asia excepting India and the Arabian Peninsula. It also includes the Atlantic Ocean east of the Mid-Atlantic Rise. The F plate includes Africa and the South Atlantic east of that rise, but not the Arabian Peninsula.

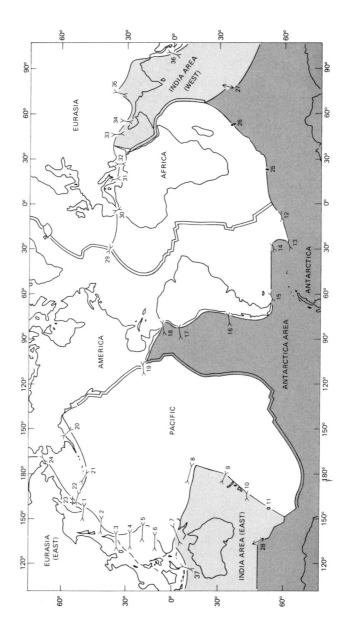

Figure 12-3. Six basic plates cover the earth, in the analysis by Le Pichon. Here, starting at left, are *(E)* Eurasian plate (Pacific end); *(I)* Indian plate (Pacific end, including Australia); *(P)* Pacific plate; *(T)* Antarctic plate (extending north alongside South and Central America); *(A)* American plate (including both South and North America); *(F)* African plate. Above the latter is the western part of the Eurasian plate, *E;* and east of it is the western part of the Indian plate, *I*.

A double line indicates a plate boundary, usually a ridge or rise, whose rate of separation was known. A single line indicates a plate boundary whose rate of separation had to be computed from the known rates. The symbol >—< shows a site of compression or coming together of two plates, while the less frequent symbol <—> indicates a site of plate separation or pulling apart.

A line of compression can be seen running through the Mediterranean Sea, then eastward through the zone of the Himalayas, south through Indochina, and so on. The small numbers show sites for which rates of movement—either compressional or pulling apart—have been measured or computed.

This is a Mercator projection.

Reproduced by permission of Dr. Xavier Le Pichon, the *Journal of Geophysical Research,* and the Lamont Geological Observatory of Columbia University.

The *I* plate includes not only India but the Arabian Peninsula and also Australia with New Zealand to the east. The *P* plate includes all the Pacific Ocean from the Aleutian Islands in the north to the Antarctic plate *(T)* mentioned below. It has islands, but no continents. The *T* plate takes in the Antarctic continent around the South Pole and most of the ocean floor south of 50° S latitude. It also includes that important sector of the southeast Pacific Ocean which lies alongside the western coasts of South America and Central America. This East Pacific sector could, indeed, be treated as a separate plate, when more detailed data is available.

Plate pulling from plate—First, the five situations where one great plate is rotating, or pulling, away from another.

(A-F) Along the Mid-Atlantic Rise, the American plate (to the west) is separating from the African and Eurasian plates to the east. The relative rotation is about 3.7° per 10 million years. The resulting expansions of the ocean floor range from about 1.9 cm per year at 60° N latitude to 4 cm per year at 30° S latitude, and drop to about 3.1 cm per year at 50° S, between the southern tips of South America and Africa.

(I-F) Along the Indian Ocean ridge, the African plate (to the west) is turning away from the Indian plate (to the east) at the rate of about 4° per 10 million years. This produces ocean floor expansion ranging from about 2 cm per year at 19° N latitude to 6 cm per year at 43° S latitude.

(P-T) In the southeastern Pacific, along the so-called East Pacific Ridge, the Pacific plate is turning away from the Antarctic plate at the relatively large rate of 10.8° per 10 million years. This produces ocean floor expansion ranging from about 5.8 cm per year at 48° N latitude, to a high of about 12 cm per year at 17° S latitude, dropping off to about 5.6 cm per year at 65° S latitude.

The preceding three sets of figures are based on actual magnetic and geophysical measurements. Two more expansion situations have been computed from the same body of data:

(F-T) Along the southwest Indian Ocean ridge, south of Madagascar, the African plate turns away from the Antarctic plate at the rate of about 3.2° per 10 million years. This results in ocean floor expansion at rates ranging from about 3 to 5.2 cm per year.

(I-T) Along the southeast Indian Ocean ridge, almost due south of the southernmost tip of India, the Indian-Australian plate turns away from the Antarctic plate at about 5.7° per 10 million years. This results in ocean floor expansion at the comparatively rapid rate of about 11.8 cm per year.

Next follow the seven situations in which plate boundaries are pushed together, causing compression and also the trench formations into which "old" ocean floor crust is thrust and consumed down in the hot mantle of earth. All seven cases were computed by Le Pichon from the data mentioned before. Actual measurements of magnetic or other geophysical nature are lacking to show positively the extent of these motions.

(A-T) The American plate rotates relative to the Antarctic plate at 5.5° per 10 million years. The results include the compressive movements into the great trench systems on the western side of South America, and also along the western side of Central America as far north as about the latitude of Mexico City. These motions are at rates ranging from 2.6 to 6 cm per year—causing the ocean crust to be devoured under the western edge of the New World continents.

(A-E) The American plate, rotating at a rate of 3° per 10 million years relative to the northwestern corner of the Eurasian plate, pushes ocean floor in a northwesterly direction under the latter plate in a sector off the west coast of Siberia. The result: plate motions of about 1.5 cm per year in this far-off corner of the Pacific Ocean.

(A-P) The American plate revolves at about 6° per 10 million years relative to the Pacific plate. The result: crust of the northern Pacific Ocean floor is forced into the trench that lies south of the arc of the Aleutian Islands, at rates ranging from about 5.3 to 6.3 cm per year.

(E-P) The vast Pacific Ocean plate rotates relative to the Eurasian plate at about 8.2° per 10 million years. Result: substantial consumption of ocean crust in the trench system lying east of the Asian mainland and its bordering islands. The rate of crustal swallowing up ranges from 7.9 in the Kuril Trench to 9 cm per year in the Mariana Trench.

(I-P) Far south—in fact, west of Australia and New Zealand—the Pacific plate rotates relative to the India-Australia plate at the quite "rapid" rate of about 12.4° per 10 million years. Result: the

swallowing up of Pacific Ocean floor crust at 10.7 cm per year in the New Guinea Trench and at smaller rates in the trenches named for North Tonga, South Kermadec, and South New Zealand. Finally, even further south, near Macquarie Island, the effect is reversed and instead of small compression there is actual separation, at the moderate rate of about 1 cm per year.

This is the only instance in which the mutual rotations produce both compression *and* tensions along the boundary between a single pair of plates.

(E-I) The India-Australia plate pushes northward against the Eurasian plate all the way from Asia Minor at the eastern end of the Mediterranean to the eastern end of India itself. This results from a relative rotation of about 5.5° per 10 million years between the two plates. The compression rates vary from about 4.3 cm per year in Turkey to 6 cm per year in the West Java Trench.

This is the same compression that continues the mountain-building that raised the awesome Himalayas. In Tibet, near those mountains, the interplate compression motion seems to be about 5.6 cm per year at present.

(E-F) The African plate, rotating relative to the Eurasian plate at about 2.5° per 10 million years, produces compressions ranging from about 2.6 cm per year near the island of Crete to 1.5 cm per year off the Azores Islands in the Atlantic, some 25° of longitude west of Gibraltar.

Summing up total tectonics—Whether the earth's surface motions are analyzed by means of six, twelve, or twenty separate tectonic plates, the same essential principles prevail:

• The earth is clad in a movable armor of relatively rigid tectonic plates, which are topped by continental or ocean floor crust and supported by the topmost or lithospheric layer of the mantle under the Moho, the boundary between mantle and exterior crusts.

- The plates interact, principally along their margins, as they are shifted slowly with respect to one another.
- The plate-to-plate interactions include the separation of plates into new plates, and even the "welding" of one plate onto another.
- The interactions result also in volcanic actions, outpourings of lava and hot vapor or liquids, and incessant earthquakes at various sites and depths.
- The plates do not propel themselves. They are carried along by convective currents moving in the asthenosphere, a mobile zone of the mantle, at some distance below the Moho.
- The energy supply for the convective currents is earth's interior heat—or rather, the flow of that heat, impelled by the higher internal temperatures, and moving toward the lower surface temperatures. The radioactive portions of the earth's mantle materials are continuing sources of release of heat energy.
- It is possible, though not completely certain, that such radioactive heat releases in the mantle alone produce sufficient power to account for all tectonic motions and effects now taking place on earth's surface.

It is certain to a high degree of probability that the energies primarily responsible for the continuing changes of the earth's exterior do originate deep within and flow ever outward. Thus arise the vast slow currents, far under our feet, on which are carried along what Julia Carney in a well-known verse once called "the mighty ocean and the pleasant land."

Almost imperceptible though they seem to us short-lived mortals, these movements cover vast distances, for—as the same verse adds—"the little minutes, humble though they be,/ Make the mighty ages of eternity."

13. Patterns and Processes

In outlining the picture that global tectonics offers of the processes taking place on earth at present, we do not wish to forget the great historical epic that first fascinated Wegener and others—namely, the vast migrations of continents during the previous couple of hundred million years.

Recent reconstructions of those migrations have been far more detailed than any Wegener could offer. A prime example is this series of five maps prepared by Robert S. Dietz and John C. Holden, geologists with the U.S. Environmental Services Administration and close students of tectonic theories (Figures 13-1 through 13-5). Such maps offer answers to questions of "where on earth" the major continental land masses were and how they arrived where they are now.

The Dietz-Holden sequence is noteworthy in supplying views of the entire globe, marked off with the present latitudes and longitudes. On these maps the contours of the continental masses may seem a little swollen. The reason is that each continent has been enlarged to include its continental shelves down to a depth of about 1,800 m (6,000 ft) below sea level. These fringes truly belong to the continental crusts, not to the ocean floor regions, which lie beyond and below.

Dietz (Figure 13-6) was one of the geophysicists who, during the early 1960s, independently proposed the concept of ocean-floor spreading, now so fundamental to global tectonics. Another was the

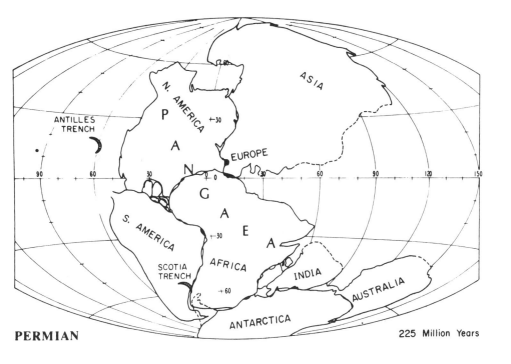

Figure 13-1. Pangaea, the universal land mass, at the end of the Permian period, 225 million years ago, as reconstructed by Dietz and Holden. The present names of the continents appear here merely to show "what became what" later on.

The dashed lines indicate boundaries whose shapes are rather uncertain. Two fixed "benchmarks" are located in terms of latitude and longitude where they are today: at left the Antilles Ocean Trench, and in the south, near the eastern side of the South American area, the Scotia Trench, near 60° S latitude.

This great single ocean is sometimes called the Tethys Sea. The projection used here spreads the entire surface of earth into a single plane, with far less distortion than the common Mercator projection.

This and the following four maps by permission of the U.S. Environmental Science Services Administration, Department of Commerce.

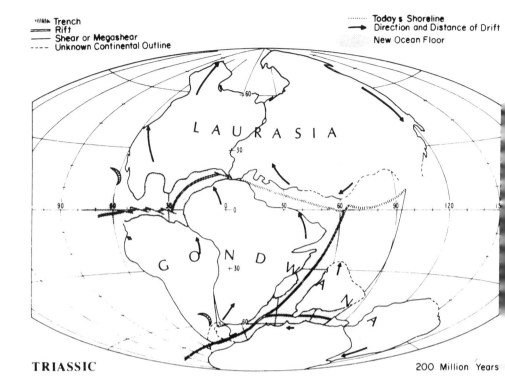

Figure 13-2. The great breakup begins. Pangaea is rifted into Laurasia (to the north) and the three segments of Gondwana (south). This shows the earth as it may have appeared in the mid-Triassic period, some 200 million years ago.

The key below gives symbols used in this and following maps. Hatched bands are deep trenches where compressional forces cause swallowing up of ocean floor crusts. Doubled lines are the ridges, or rises, where tensional forces cause new crust to emerge and spread. Dark tones indicate new areas of ocean floor crust. Dotted lines indicate the shorelines of today. Solid single lines show major faults or fractures, sometimes called shears or megashears. Dashed lines indicate unknown continental outlines of the past.

Dark arrows indicate direction and relative distance of drift during the period whose outcome is shown by the map. Note that the subcontinent of India, just above the Antarctic, is heading northward!

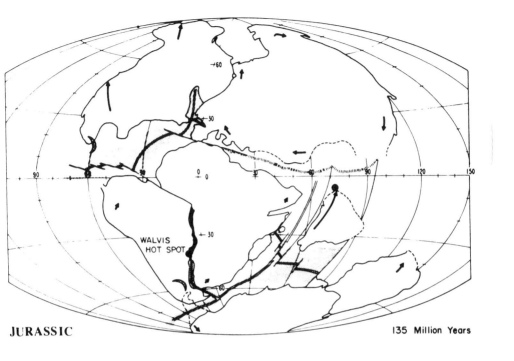

Figure 13-3. Continued continental scattering by the time of the late Jurassic period, 135 million years ago.

Two new "fixed features" have been added: first, the so-called Walvis hot spot, an active thermal center, here shown in the common boundary between South America (to the left) and Africa (right); and, second, another "hot spot," slightly south of the equator, and here just north of India, whose further progress northward is easily apparent. These two hot spots are seen also in following maps.

Note that North America here has begun to split away from Eurasia.

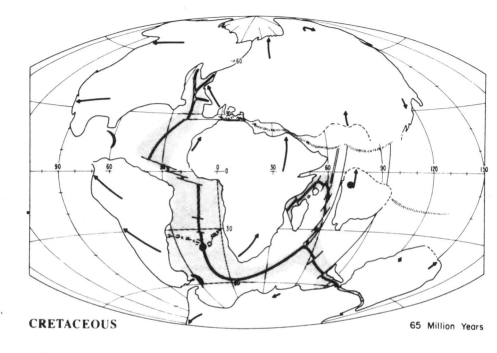

Figure 13-4. The great dispersion a stage later, at the end of the Cretaceous period, 65 million years ago. Here, the South Atlantic is well defined, and the North Atlantic has made a pronounced start.

Note the doubled or branching rift, which is beginning to separate Greenland from Europe to the east and from North America to the west.

India has begun to cross the equator, still northward bound.

At lower right, Australia has been rifted loose from Antarctica and is moving northward.

The great ridge-rift system here loops entirely around Africa to the west, south, and most of the east.

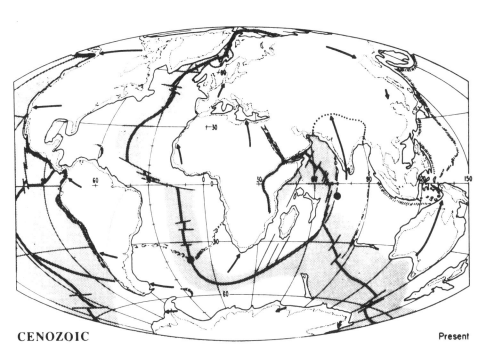

CENOZOIC Present

Figure 13-5. Today's scattered and characteristic continents. The dotted interior outlines show present coastlines. The solid lines beyond them show the extent of the continental shelves attached to the continents.

Note the extent of the ocean-floor spreading (shaded) from the Cenozoic period to the present. Note also the marked rift line plunging southwest and then south into Africa from the southern end of the present Red Sea.

Here India and Australia have completed remarkable northward flights at relatively rapid rates of about 10 cm (4 in) per year. India, in fact, has encountered and underthrust the Asian plate, thus giving rise to the great Himalaya mountains and highlands.

The great ridge-rift system now loops and curves around the world, much like the seams in a baseball. Extensive deep-ocean trenches, here shown at work, surround much of the shrinking Pacific Ocean, and other deep trenches also are active elsewhere.

late H. H. Hess of Princeton University. He called the idea "geopoetry," so imaginative and exciting did he find it.

Figure 13-6. Dr. Robert S. Dietz, marine geologist and geophysicist, co-founder of the concept of ocean-floor spreading, and one of many scientists active in developing and extending the ideas of global tectonics.
Photograph courtesy of U.S. Environmental Science Services Administration.

The name "geo*drama*" might well be applied to today's tectonic interpretations and speculations. Most dramatic, perhaps, are the concepts of the enormous forces at work and the vast periods of time during which they have operated. In exploring these dramatic possibilities we shall ask, and seek to answer, two further questions: (1) Can earth conceivably provide the power required for the movements of entire continents and the opening of vast oceans on earth's surface? And (2) What physical processes could link the sources of such power with the actual movements?

Power enough?—Earlier, this book noted that measurements around the world indicate a total outward flow or current of heat to the surface at the rate of 32 million MW (32×10^{12} W). This gives some idea of the scale on which the earth operates as a heat engine.

Typical surface temperatures of earth are significant. The ocean floor, at considerable depth, remains at about 0° C, for the sun's heating effect does not penetrate through so much seawater. On the continents, one must dig some feet or a few meters down in order to get past the temperature variations caused by glaring sun or cold snow on top of the ground. Taken as a whole and on a year-round basis, the average temperature at, or quite near, earth's surface is about 14° C, 57.2° F, or 287 K on the absolute Kelvin scale.

Most of this temperature can be attributed to the warming effects of the sun's light and heat radiation. However, a part of it appears to result from added heat energy flowing to the surface from below. Average temperatures just under land surfaces are found to lie between about 1° C and 5° C higher than the averages for the air just above. These differences are most marked where little rain falls, since such moisture tends to equalize air and ground temperatures.

The power of the outward flow of heat energy from inside the earth is nevertheless tiny when compared with the power of the solar energy that reaches the earth constantly. That great flood of radiation has a power of 173,000 million MW. About 30 percent of it is reflected back into space, much as the moon reflects sunlight to us. The remaining 121,000 million MW, which does influence the earth, is about 3,800 times as powerful as the heat current which flows from inside to the surface!

In fact, even the relatively tiny fraction of the sun's power that the plants of land and sea absorb for their photosynthesis—about 40 million MW—is greater than the apparent power of the heat flow from inside. And the latter is only one-tenth of the magnitude of the power constantly operating in earth's winds, waves, currents, and other activities of atmosphere and oceans.

Small wonder, then, that in the mid-1940s, before global tectonics had emerged with all its geodrama, the *Handbook of Meteorology* stated flatly, "The minute amounts of heat flowing from the interior of the earth outward . . . are negligible."

Yet today the most competent geophysicists confidently reckon on the internal heat supply of earth as the source for the power that not only forms but moves mountains, continents, and vast oceans. How is this possible?

Slow and easy does it—For one thing, these motions are extremely gradual. An airplane typically moves at about 1,000 km per hour, an automobile at about 100 or 200 km per hour, but a drifting continent or a spreading ocean floor at 1 to 10 m per century! The speeds are so slow that though the total energies are large, the rates of power conversion remain within the possibilities of the interior heat flow, as far as can now be estimated.

We have, indeed, some evidence that even smaller amounts of power suffice for enormous geologic effects on and in our earth. Seismologists have estimated the flow of power represented by all the earthquakes, great and small, constantly taking place on earth. The largest of such estimates is 10^{19} joule per year, equivalent to about 3.2 million MW of power. This is only a tenth of the power of the heat flow from inside earth to its surface.

Specialists also have estimated the power of the total flow of heat that takes place through such special transports of matter as hot lava in volcanoes, thermal hot springs, and geysers. For all this a total of about 0.3 million MW is found sufficient. It is less than 1 percent of the power of the overall heat flow from inside the earth.

One further yardstick may be applied. The tides produced by the attractions of the moon and sun result in friction as the oceans of earth are forced through channels and straits. Result: the rate of the earth's spin slowly lessens. Once the earth turned in 4 hours, now it needs 24, and each million years the average day grows some 16 to 17 seconds longer. In 4 million years the day lengthens by more than 1 minute, and in 240 million by more than 1 hour.

Yet the frictional power thus retarding the earth's spin is only some 1.1 million MW. If so small a fraction of the heat power flowing from within the earth is sufficient to retard the turning motion of the whole earth, then 32 million MW or even a third or quarter of it can surely shift the principal features on the surface of the earth!

The outer crust or rind of earth, as seen by global tectonics, has been said to resemble a beat-up old baseball whose outer cover has worked loose from the insides (Figure 13-7). Truly, earth's plate-crusts move, or are moved, independently of the great bulk of the mantle below them!

How much from within the mantle?—This independence of the great ball's covers from its interior is the more striking, because the great convections that move the plates bearing continents and oceans almost certainly are powered primarily by that part of the radioactivity which takes place in the mantle *below the asthenosphere.*

We can as yet but offer guesses just how powerful that portion of the total radioactivity may be. However, the best available recent estimates suggest that the low rate of radioactivity in the ultrabasic rock of the mantle is compensated by the vastly greater volume of mantle material as compared with the volume of the continental and oceanic crusts, in which radioactivity is much more intense.

It appears, in fact, that the total power of radioactive heat released

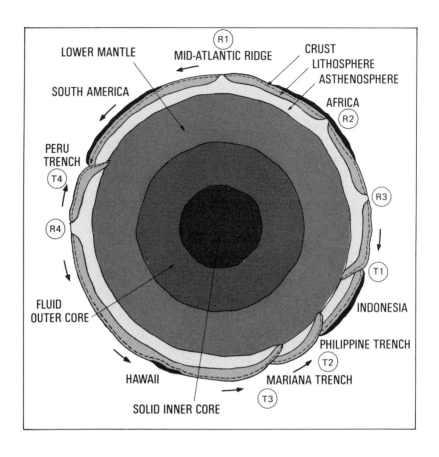

Figure 13-7. Earth's cover—its crustal plate system—has worked loose from its core, like the cover of a battered old baseball! This cross section of the globe suggests special sites where the cover sections shift and slide over the interior substance.

Starting with the Mid-Atlantic Ridge or Rise at the top, we find a total of four ridges (numbered R1 to R4), each of which is formed by the pulling apart of the great plates below, in the conveyor belt action of the asthenosphere layer underneath the crustal lithosphere. Under eastern Africa a rift (R2) is even now tending to split that great continent. Then east of Africa and south of India, the ridge through the Indian Ocean is indicated by R3.

All around the great Pacific region (lower half of the sphere) no rifts appear until the East Pacific Ridge (R4) at left. It lies in a general southerly direction from the Gulf of California and is helping to split Lower California from the Mexican mainland.

Four trenches are labeled T1 to T4. The first, at T1, lies west of Indonesia, and is the site where crust of the Indian Ocean is swallowed up as it moves eastward. The next, labeled T2, lies east of the Philippines, and is the site where Pacific ocean floor crust, moving westward, is swallowed up under the eastern edge of the Asian plate. Next, labeled T3, is the Mariana Trench, lying north and east of the Philippine Trench. Other Pacific ocean floor crust, moving westward, is swallowed up in this deep Mariana Trench.

No further trench sites are shown except the great Peru Trench (T4), more than a third of the way around the world. Here, ocean floor crust which was generated in the East Pacific Rise is swallowed up as it moves eastward. Such swallowing up of ocean floor crust takes place along almost all the western sides of South America and Central America.

The trenches result from the head-on meetings of the plates pressed against each other by the convection conveyor belt system below them. The ridge-rifts are the result of the separations of the plates, moved away from each other by another phase of the same global convection system working from below.

Modified from an illustration in *Horizons,* magazine of the Amoco Production Co., Tulsa, Oklahoma.

all through the earth is very nearly 34 million MW—or quite close to the estimated heat flow to the surface—and roughly 35 percent of this radioactive heat release takes place within the mantle, rather than in the granite (sial) of the continental crusts or the basalt (sima) of the ocean floor crusts.

Radioactivity is not the only means by which temperatures within the mantle have been raised to their present levels. Through the eons since the earth was formed its interior was heated first by enormous compression, due to the sheer weight of overlying layers. Then, at some stage, began interior processes which changed a relatively uniform earth into the present segregated or layered earth, with a dense core of iron and nickel, a less dense mantle, and, least dense of all, the thin shells of ocean floor and continental crusts.

Gravitational energy was released, doubtless resulting in higher internal temperatures, as the heavy iron separated out and sank down to form the core, while the lightest components were lifted to the surface, where they form the "froth" that we know as continents.

With this enormous separation process, the radioactive atoms, having greater affinity for compounds within the froth or slag of earth, became increasingly concentrated in the continental granites, less concentrated in the oceanic basalts, and least concentrated in the dense ultrabasic mantle rock.

A model of how these separations worked is offered here (Figure 13-8). It pictures a stage at which only two enormous continental masses—Laurasia and Gondwana or Gondwanaland—had been accumulated on opposite sides of the globe, each as far as possible from the equatorial mid-ocean rise, and hence centered over the convective "downdraft." The great convective currents sweeping through the mantle are clearly indicated.

Such long-time processes of separation and accumulation account well for the marked differences in average radioactive heat release in like volumes of four different kinds of substance in the earth today:

Type of substance	*Average power of radioactive heat release per cubic kilometer (km^3)*
Granitic (sial) rock, as in continental crust	2,310 watt
Sediments, on top of continents, etc.	840 watt
Basaltic (sima) rock, as in ocean floor crust	460 watt
Ultrabasic, mantle-type rock	13 watt

These figures are necessarily approximate, but there is no doubt of the extent of the major differences.

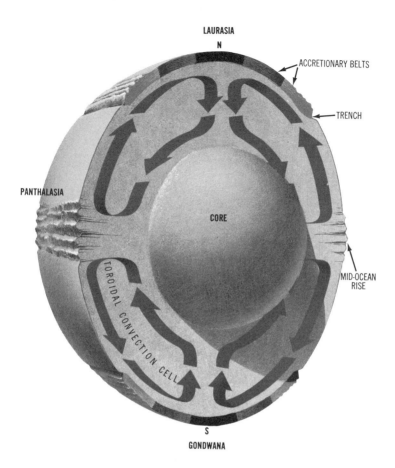

Figure 13-8. The earth's crust and mantle may have looked like this long ago when two-cell convection was at work. The lower convection cell has accumulated the continental mass called Gondwana at the bottom (marked S for south, here). The upper cell has accumulated the continental mass of Laurasia at the top (N).

The "accretionary belts" show how the conveyor belt movement has carried light granitic (sial) matter to the continental margins, gradually extending them. The two continental masses are each centered above descending convectional flow. And at edges of the continental blocks are the deep trenches into which the ocean floor crust is carried as it underthrusts the continental plates.

Over the ascending or updraft sites of convection, the mid-ocean rise has formed. In this case the great rise would circle the globe just about at the equator.

The name Panthalasia, or more properly Panthalassa, has been given to the single great ocean that blankets the earth between the two antipodal continental masses.

Illustration by Jim Schick, from *ESSA,* April, 1970, magazine of the Environmental Science Services Administration, U.S. Dept. of Commerce.

As precise measurements of surface heat flow began to be made at various land locations, a prevalent average equivalent to 63 kW per square km was found. Knowing that the ocean floor basalts were so much less radioactive than the continental granites, geophysicists expected that heat flows at the top of the ocean floor would be far less than those near the top of the continental crust. Indeed, it is possible to figure that within 100 km below a typical square meter of continental surface, about 10 times as much radioactivity is at work as in a like depth under a typical square meter of ocean floor.

Even if the calculations include the situation all the way from the surface down to the bottom of the mantle (just above the earth's core), the radioactivity under continents should be anywhere from two to three times as powerful as the radioactivity under ocean floors.

Once again, expectations were upset by experience—which is to say, by accurately measured testing. The needlelike thermal probes inserted into many parts of the ocean floor soon showed that the average heat flow here was just about as high as on the continents! This lack of difference was one of the most stimulating results in experimental geophysics. It became a landmark like Sherlock Holmes's famous clue of the barking dog. What was the clue? On the night of the crime the dog had *not* barked. Science, like detection, advances by meaningful negatives as well as powerful positive results!

How could this approximate equality of heat flow come about? It must be that additional heat was somehow reaching the ocean floors from elsewhere within the earth—that is, from the great mantle which underlies ocean floor and continental crusts alike, down to the core nearly 2,000 km below.

Unquestionably the lower mantle matter was at far higher temperatures than any of the crusts. But how could heat be flowing in sufficient amounts to the ocean crusts? Ordinary heat conduction in a static mantle could not suffice. A different, more rapid and effective method of heat transfer must be increasing the flow to and through the ocean floor crusts.

Only one conceivable method fits the possibilities. Its name is *convection.* It has been described earlier in this book. Our seemingly solid earth is actually transferring vast amounts of interior heat toward the surface in much the same way that the blower on a small electric heater propels quantities of heated air from the glowing wires into the room that needs greater warmth!

Curious kinds of convection—The uppermost mantle, just under the Moho boundary, is known to lie higher below the oceans than below the continents. Thus, if under oceans currents of hotter than usual mantle material rise to, or near, the Moho boundary, the resulting additional heat flow through the ocean floor crust should be especially marked. The shorter the distance the heat has to travel, the less resistance it must overcome.

We have seen, further, that as much as 35 percent of all earth's radioactive heating may take place in the mantle rather than in the crusts. There is no doubt that the really high temperatures inside the earth are located at depths below rather than above 100 km down.

We have seen that continents are slowly moved *away from* the mid-ocean rises, where the great plates separate, and that hot mantle material oozes up into the gap, forming new ocean crust which then spreads both ways from the rise. The continents then are moved away from these sites where much interior heat emerges, and toward the deep trenches, which are sites where heat flow from the interior is lower than average.

Continents are thus propelled away from the "hot spots" of our earth, and the crusts of the ocean floors are formed at, and creep outward from, zones marked by greater than ordinary flow of heat from the interior.

The underlying convective movements cause new ocean crust to emerge *from* zones where the heat flow from below is especially high. Meanwhile, continents are pushed or pulled *away* from such zones. The enormous role played by convection is revealed by recent calculations of Professor H. Takeuchi of Tokyo University, indicating that convection transports *half* of the total heat flow to the surface, the rest moving by means of ordinary heat conduction. Thus convection continually carries up some 16 million MW of heat power, besides all the mechanical power represented by the motions of the masses of matter within the mantle.

Yet how, in mechanical terms, are these mantle and asthenosphere movements converted into motions of the great tectonic plates on earth's surface? Are the forces supplied simply by the upflowing of hot mantle material at the rifts along the ocean rises? That process seems to be a result rather than the cause of it all. Or does the sheer weight of the ocean floor crusts, as they slide down into the deep-ocean trenches, pull the rest of the plate along in that same direction? This has been suggested, but seems hard to picture.

Much evidence seems to favor a quite different and more concentrated mechanism. Principal architect of this model is W. Jason Morgan of Princeton University, with Kenneth S. Deffeyes of that institution as an active expounder. The following is a brief description of Morgan's bold model.

Plumes on parade—Scattered about the globe in significant locations are about twenty different "hot spots." They are identifiable by threefold heightenings: (1) heightened (higher than average) heat flow at the surface; (2) heightened (higher than average) gravitational intensity; and (3) heightened (higher than the surroundings) surface elevation. Thus, though small—hardly more than 400 or 500 km in diameter is the estimate for the vertical interior "pipes" that create them—they are outstanding, thermally, gravitationally, and physically.

Most of these hot spots are found at or near oceanic rises, where two great tectonic plates are being pulled apart (Figure 13-9). A few, in fact, are at or near *triple* junctions, where three (not two) different plates come together. Then each of the three plates is found to be moving in a different direction from the others, as if some mantle currents were emerging from the hot spot and flowing in all directions away from its center, carrying the overlying plates along.

Each of the hot spots is assumed to be at the top of a long narrow convection updraft—a sort of pipe or *plume* through which hot matter moves up from high-temperature regions below. Indeed, it is suggested that the average rate of rise is rather rapid—as much as 10 cm per year—and that marked temperature differences—as much as 300° C—exist between the ascending substances within the pipe and those at a similar depth in the mantle outside the pipe.

Instead of shallow mantle convection "cells," which some geophysicists have suggested, Morgan pictures the plumes or pipes operating from the very depths—perhaps even from the base of the mantle, near the central iron core, more than 2,800 km below earth's surface. The entire upward journey in one of these strange tectonic

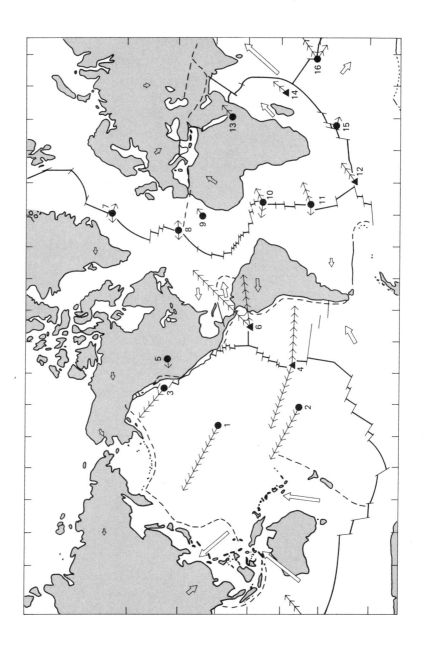

Figure 13-9. Hot spots dot the world and are the power sources from which the great tectonic plates are moved, as indicated in this global map on the Mercator projection. Here some 16 of an estimated 20 hot spots or convection "plumes" are indicated by numbers.

Long arrowlike indicators leading from the hot spots show the directions in which the plates are being pushed, relative to those hot spots. Small open arrows, on the other hand, indicate relative speeds as well as directions in which plates are moving with respect to the underlying mantle of the earth.

Four hot spots are marked by triangles. These are at or near triple junctions and are believed to provide the underground currents that move each of the three plates nearby. Those four are identified by the letter (T) in the following key, which gives the approximate geographic locations.

Three hot spots are under plates not yet split or broken apart. They are identified by the letter (C), for "covered," in the key.

The remaining hot spots are at or near ocean rises (ridges) and are believed to supply the currents moving apart the plates on either side of the ridge. These are identified by the letter (R), for "rise."

Key to earth's leading hot spots (convective plumes):
1. Hawaii (C)
2. Gambier, in French Polynesia (C)
3. Juan de Fuca Strait (R)
4. Easter Island (T)
5. Yellowstone, U.S.A. (C)
6. Galapagos (T)
7. Iceland (R)
8. Azores (R)
9. Canary Islands (C) or (R)
10. St. Helena (R)
11. Tristan da Cunha (R)
12. Bouvet (T)
13. Afar, near Djibouti (R)
14. Reunion Island (T)
15. Prince Edward (R)
16. French Islands of Kerguelen, Amsterdam, and St. Paul (R)

These identifications and analyses will very likely be modified as well as expanded in number and detail later on.

Map data from W. Jason Morgan, Princeton University.

elevators might be completed, from near the core to near the surface, in the geologically brief time of about 25 to 30 million years.

The trip up, however, is the beginning rather than the end of the important convective process. As the rising hot matter reaches the particular upper mantle zone, the asthenosphere, in which fluid rather than solid conditions prevail, a spreading out takes place. This spreading may be expedited or enforced by a so-called temperature inversion within the asthenosphere zone. In any case, over the semi-liquid asthenosphere lies the relatively rigid top mantle layer: the lithosphere, on which the ocean floor and continental crusts rest. As the lithosphere is moved, so are the crusts moved.

So far we have pictured only the updrafts, concentrated and rather swift. But what goes up from the interior must eventually sink down again, else the earth would be an expanding hollow sphere! The Morgan model pictures the hot mantle material, after spreading laterally in the asthenosphere, cooling and settling back again toward the bottom—but settling with extreme slowness, at rates as low as 1 mm per year.

If an imaginary concentric shell could be placed within the earth below the asthenosphere, about 1 percent of its total area would be penetrated by the pipes or plumes bringing up the hot matter convectively, while the other 99 percent would be areas through which cooler matter settled down again almost imperceptibly.

A complete round trip—a full cycle—from lower mantle to the asthenosphere, through the asthenosphere, and then down again to the depths—might be completed in about 1.5 billion years, or about one-third of the earth's present age! The typical travels of a morsel of mantle matter may well include: (a) rising via a plume about 1 percent of the time, (b) spreading outward in the asthenosphere 2 or 3 percent of the time, and (c) slowly sinking the rest of the time. The sinking should, of course, be accompanied by reheating, until the bit of matter, once again too hot and expanded to stay put, begins a new upward movement through one of the plumes.

Each of the limited number of hot spots serves as a driving center, *away from* which the overlying plates and the underlying lithosphere are slowly carried.

The ocean floor and continental crusts are too thin to be really rigid and strong, but supported by underlying lithosphere they move as rather rigid and permanent units, propelled from these thrusting centers or hot spots at or near their margins.

Hot spots may operate under the interior or even the center of a great plate. In the first stages they cause the plate to bulge upward above them, and to crack. Through those cracks heat, gases, and lava rise, and volcanoes or volcanic islands begin to appear on the surface above.

Eventually the strains above may become so great that the plate is split apart and increasingly separated. Such was the origin of the great divorce between Africa and South America, with the subsequent formation of the South Atlantic; and between Eurasia and North America, with the formation of the North Atlantic.

What has become of the thermal pipes and hot spots that accomplished that tremendous splitting and ocean-building? They are still at work, along the line of the Mid-Atlantic Rise. Beginning far in the Southern Hemisphere one may find them, near Bouvet Island, then farther north near Tristan da Cunha, and near St. Helena, where Napoleon died. These hot spots apparently suffice to continue driving apart Africa to the east and South America to the west, and in the process expanding the Atlantic Ocean floor.

In the North Atlantic we find active hot spots near the Azores and at Iceland, whose thermal and volcanic nature is famous. These force centers seem able to maintain the growing separation between Europe to the east and North America to the west, with the resulting expansion in the North Atlantic Ocean floor.

The Pacific plate is marked by two great hot spots which have not yet split it apart, but have provided it with significant chains of volcanic islands. One is the Hawaiian hot spot, well north of the equator; the other is the hot spot near Gambier in French Polynesia, far south.

To Americans the most remarkable of the "covered" hot spots is that which emits the vast heat displays identified with Yellowstone in Wyoming. Here, as elsewhere, the terrain roundabout is raised to altitudes which can be accounted for geologically only by assuming that something powerful thrusts up from below.

Why so slender and swift?—The Morgan model of tectonic convection seems almost startling at first, largely because of the concentrated nature of the thermal pipes, or plumes, and the hot spots that mark their summits. Yet this is far from the only instance in nature in which great circulatory processes give rise to narrow, swift, localized currents in one phase, coupled with broad, slow settling back in another phase. Scientists of the weather and airplane pilots know well how, over hot continental lands in summer, slender, swift updrafts are born, sweeping up as high as 60,000 ft, and there building the vast billowy cirrocumulus clouds. At certain levels these updrafts will mushroom outward, spreading laterally instead of rising vertically still further.

Ocean-watchers and surfers know well what happens when huge battering surfs sweep in toward shore, building the average level of inshore water so high that a great *head* or pressure develops. The excessive inshore water "wants out" again.

The old supposition or superstition was that this water flowed outward under the incoming breakers in a broad bottom current called "the undertow." The fact is that the outflow takes the form of narrow, swift streams, visible on the surface, called *rips* or *rip currents.* They rip through the incoming breakers, and are as concentrated in their way as the thermal updrafts over hot land or the convective plumes that Morgan finds rising from deep in the earth and powering the massive movements of continents and ocean floors. The author has demonstrated the action of the surf-centered rips in his book *Surfing* (Lippincott, 1965).

From earliest origins to the present, the evidences of continental drift and related events on earth have surprised or even startled attentive observers. Geology and geophysics have a way of deflating human assumptions and unsettling petrified prejudices. Rethinking things is often trying, or even a little painful, yet it is exhilarating, too.

The poet Robinson Jeffers, previously mentioned, stated it simply: "What a pleasure it is to mix one's mind with geological/Time, or with astronomical relax it" ("Star-Swirls," 1963).

14. Applying the New Insights

Important new answers in science tend to generate new and more searching questions. This has been the case as the revolutionary new insights of global tectonics have developed during the 1960s and 1970s. Among the readers of this book, perhaps, are some who later on will help to contribute further answers.

The new concepts have already proved practically useful as well as theoretically satisfying and stimulating. Many a finding that would have been misunderstood or perhaps even overlooked ten or twenty years ago now receives new attention and interpretation.

Thus, late in 1971 the famous oceanographic research vessel, *Glomar Challenger,* returned from an expedition taking submarine rock cores from the floor of the Pacific Ocean. About 800 miles southwest of Tokyo samples were raised from more than 1,200 ft below the ocean floor, which at that spot lay 20,000 ft below the ship. The samples contained sediments formed from sea life of a type still being laid down in a narrow belt, no more than some 200 miles wide, at the earth's equator!

There, in a zone formed by interaction of ocean currents in the Northern and Southern hemispheres, the sediments are typical and rather easily identifiable. The conclusion seemed easy: the sediments had somehow been moved about 1,600 miles to the north of their point of origin. The estimated duration of the journey was at least 135 million years, corresponding to an average speed of 1.8 to 2.0 cm (0.7 or 0.8 in) per year.

The striking content of this core became one more confirmation, rather than a source of confusion, thanks to global tectonic insights. Elsewhere the expedition's scientists found the ocean floor strewn with volcanic ash sediments, identified with volcanoes along the Asian coast—and suggesting that this portion of the Pacific floor was being thrust toward and under Asia at the rate of about 10 cm (4 in) per year.

Prospecting for petroleum—Almost a decade ago, enough evidence for drift theory had accumulated to persuade some alert petroleum geologists that these same strange ideas might help them locate regions containing stores of petroleum as yet untapped.

"A comparison of the coastal regions of eastern North America, northwestern Africa, and western Europe indicated that [these] continents could indeed have been a continuous mass until late Triassic or Jurassic times." The words are those of D. E. Powell, regional geologist with the Pan American Petroleum Corporation, at Tulsa, Oklahoma.

Modern maps, however, showed no land areas that might, in those past times, have filled the gap between present-day Portugal and Newfoundland. Seeking for that missing land mass, these petroleum hunters hit on the idea that it might be the land that now lies below the Grand Banks, a large region of shallow water east of Newfoundland.

If that should prove true, they hoped the geology below the Grand Banks would resemble "the geology of western Portugal," and thus afford some new oil deposits. The result: "Pan American then launched a leasing and geophysical program on the Grand Banks."

In 1971 the Standard Oil Company of Indiana reported heavy commitments in this search. Massive special equipment includes a huge three-legged driller big enough to dwarf two-story dwellings.

Clues acquired during drilling operations in the Grand Banks bottom would have delighted Alfred Wegener, for they merge geology, geophysics, and micropaleontology. Here (Figure 14-1) is the fossil remnant of a dinoflagellate that lived about 86 million years ago before its skeleton drifted down to become part of the sediments under the Grand Banks. Here also is another dinoflagellate, clearly of the same species, but found in sediments at Folkestone, in Kent, England. It lived at about the same time. Both these tiny creatures clearly descended from ancestors who swam in seas before the strains, rifts, and drifts that formed the present Atlantic Ocean.

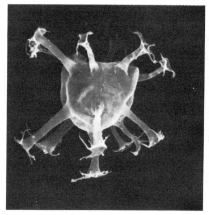 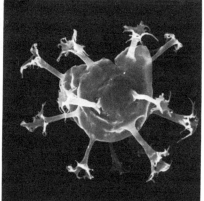

Figure 14-1. The fossil shown at left, so small that ten of them would fit into the period at the end of this sentence, was taken from a rock core from the bottom of the Grand Banks, off Newfoundland. The name of the creature whose skeleton is shown here is *Oligosphaeridium complex;* it lived between 70 and 86 million years ago. Its identity with the fossil from England (right) is strikingly clear. Such micropaleontological likenesses are among the many evidences for the reality of far-reaching continental drifts.

From *SpaN,* magazine of the Standard Oil Company of Indiana.

The new global tectonics links widely separated lands and eras. It brings together separate scientific specialties in the very way that Alfred Wegener once urged, and as Francis Bacon himself recommended when he wrote that the limits of the individual sciences should be set merely for convenience, and not as "sections to divide and separate."

J. Tuzo Wilson (Figure 14-2), a Canadian who has contributed much to the new tectonics, has welcomed the opportunities that its concepts provide to "integrate geology and geophysics." Such integration and mutual stimulus is at work now as never before. It marks the theoretical and also the applied aspects of global tectonics (Figure 14-3).

Figure 14-2. J. Tuzo Wilson of the University of Toronto, Ontario, Canada.
Photography by Karsh, Ottawa.

Figure 14-3. In action—the integration of geophysics and geology that Alfred Wegener advocated. Before a modern tectonic map, two members of a research team study the vast plate motions revealed by current continental drift analysis. David Graham (left) and Peter Link, senior research scientists of Amoco Production Co., Tulsa, Oklahoma, are examining deep ocean basins and their sediments, in search of formations promising to bear oil.

Graham points to a string of arrows showing the general direction of the great plate bearing the Pacific Ocean, from the East Pacific Rise (west of South America) to the deep-ocean trenches off Asia. The trenches appear as heavy cross-hatched lines. Mid-ocean rises, on the other hand, are shown by plain heavy lines.

Just behind Link's nose can be seen the great curve of ocean rises that loops around south of Africa, then veers northward as the Mid-Atlantic Rise, all the way up to and through Iceland. A curving line of arrows through northern Canada indicates how that part of North America rotates. Other arrows show conflicting plate motions in northern South America, in southern North America, and in the Caribbean region. A separate small Caribbean plate is here shown moving toward the east, while both the North American plate above it and the South American below it move westward. (This plate analysis is more detailed than that used by Le Pichon, who simplified the situation to just six major plates.)

Far east of the southern tip of South America, a small cross-hatched line suggests the presence of a short, curved deep-ocean trench. It lies just about where the so-called South Sandwich Islands appear on a globe of the world.

Photograph courtesy of Amoco Production Co., Tulsa, Oklahoma.

The results are exciting. "It's a revolution. We are riding the crest of a breaking wave," said a sea-oriented scientist of the earth, Melvin Peterson of the Scripps Institution of Oceanography at La Jolla, California.

To specialists and to interested laymen also, this revolution in the earth sciences affords amazing new insights and understandings. Our earth is revealed as a complex of continuing processes, rather than a storehouse or museum of events completed long ago.

This planet did not undergo a complete and final "creation" at any time during the eons since it was assembled. It has, instead, continually been creating and recreating itself. The earth that is the only home and refuge for all humanity is dynamic, mobile, and thus more challenging to the imaginations of its human passengers.

A link with life—Some threescore years ago, Alfred Wegener enlisted the aid of some of the life sciences in amassing evidence for his theories of continental drift. These included paleontology, zoogeography, and phytogeography. Now it is becoming apparent that global tectonics, the energetic offspring of the drift theory, should provide new insights for the life sciences.

This has been suggested in a brief but beautiful summary by William E. Benson, head of the Earth Sciences section of the National

Science Foundation. Dr. Benson kindly permits the inclusion here of this statement.

Our Mobile Earth

The history of any science is punctuated by periods of rapid advance separated by times of relative stability and consolidation of ideas. Almost without exception, the rapid advances are the result of a new and simplifying theory that unifies thinking and provides a new framework for the interpretation of observational and experimental data. When the new theory is so far-reaching as to affect the whole field, we commonly speak of the consequences as a revolution. And a revolution in one field of science inevitably opens new vistas and opportunities in related fields.

During the past decade, geology has undergone a major revolution, the second in its long history.

The first occurred in the late eighteenth and early nineteenth century and was occasioned by the development of Hutton's* principle of *uniformitarianism*. First published in 1788, the principle gained credence slowly until it was incorporated into the famous works of Charles Lyell** in the early nineteenth century. Commonly stated as "the present is the key to the past," this principle maintains that the features we see on earth today have originated by processes that we also see operating today. Because most natural processes are geologically slow acting, the earth therefore must be millions or even billions of years old.

Before Hutton's enunciation, ideas about the physical features of the earth had been constrained by a venerable (and unproven)

---

*James Hutton (1726–1797), Scottish geologist, author of *Theory of the Earth* (1795).

**Charles Lyell (1797–1875), outstanding British geologist, author of *The Principles of Geology*, to which he gave the following significant subtitle: *An attempt to explain the former changes of the earth's surface by references to causes now in operation.*

belief that the earth was no more than a few thousand years old. Given so short a history, the earth's physical features must have originated as the result of extraordinary forces or cataclysmic events. But with the acceptance of the principle of uniformitarianism it was suddenly a new ball game—there was plenty of time for all of the earth's features to have developed slowly and in an orderly fashion.

The principle of uniformitarianism (and the concept of "geologic time") was probably as important in biology as in geology, for it was this principle that enabled Charles Darwin to construct his theory of evolution on a scientific foundation. Without the principle of uniformitarianism, the history of life of the past could have had a thousand interpretations based on physical conditions entirely dissimilar to those of the present.

The modern revolution in geology has not yet been labeled with a single title but is commonly summed up in terms such as *sea-floor spreading, global tectonics,* and *plate tectonics.* The aggregate for the new concept derives from several subfields of geology and geophysics, some of them—such as paleomagnetism and marine geology—relatively new, some of them—such as seismology, gravity, and paleontology—relatively old. The cement for the aggregate is the revival of the old and controversial idea of *continental drift* as modified by the new theory of sea-floor spreading.

By whatever name the new theory is called it has provided earth scientists for the first time with a unifying concept of global structure and composition, and a new framework in which to set detailed geologic studies.

Stated briefly, the new concept holds that oceanic crust is being generated constantly along the axes of the mid-ocean ridges. Hot molten material rises from the underlying mantle, cools, and is forced aside by new material coming from the depths. Thus the ocean floor is in movement, the principal direction being at right angles to and away from the center of the mid-ocean ridges.

Ultimately the oceanic crust moves to an area such as the deep sea trenches where it is drawn down and reabsorbed into the mantle.

Because the earth's crust is not infinitely strong, the whole crust does not move as a single unit, but is broken up into major plates. Some move parallel to each other (as at the spreading centers), and some collide.

As the plates move they carry not only the upper part of the oceanic crust but also the continents. For example, the Americas are now thought to have been joined to Europe and Africa until about 200 million years ago, when a rift or spreading center split them along what is now the line of the Mid-Atlantic Ridge. Since that time, the new and old worlds have been moving apart at a rate of 1 to 2 centimeters per year.

Even as the concept of uniformitarianism in geologic time has profound repercussions in biology, so the new global tectonics opens up a new era of study in the geographic patterns of distribution of species and the evolutionary process. In addition, some of the implications of paleomagnetism have bearings on the mechanism of evolution.

Figure 14-4. *Farewell Forever*, or *A Ruptured Romance of Long Ago* . . . Could continental cleavage and drift have cut like this across the course of true love? This geo-fantasy, cartooned by William C. Littlejohn, is based on a notion borrowed, with grateful acknowledgement, from Dr. Robert S. Dietz, geophysicist.

Index

Page numbers in italics refer to illustrations.

accretionary belts, *161*
Afar Triangle, 82
Africa, 17-20, 22, *25,* 27, 37-39, 46, 66, 68-69, *71,* 71-72, 82, 96, 121, *125,* 137, 141, 144, *151-153, 158,* 169, 173, *176,* 180; coast of, 37; East, 81; South, 68; West, 29
African plate, 141, *143,* 144, 146
aftershocks. *See* earthquakes
Alaska, *27,* 111, 130; Alaskan peninsula, 130
Aleutian Islands, *27,* 118, 130, 143, 145
Aleutian Trench, 130-131
Almond, M., 114
Alps, *25*
America(s), 18, 22, 27, 96, 120, 180
American plate, 84, 141, *143,* 144-145
Ampferer, Otto, 96
Andean Islands, *26*
Andes, 40, 79
Antarctica, 39, 41, 46, 124
Antarctic continent, *27,* 143, *152*
Antarctic plate, 141, *143,* 143-145
Antilles: Greater, 42; Lesser, 42
Antilles Ocean Trench, *149*
Arabia, 82
Arabian Peninsula, 141, *143*
Arctic, 31, 37, 50; Circle, 120
Arctic Ocean, 45
Argand, E., 69, *115*
Arizona, 117
Asia, *26-27,* 38-41, 46, 69, *88,* 92, 97, *115,* 116, 121, 131, 136, 138-139, 141, 145, 146, 173, *176*
Asian plate, *153, 159*

asthenosphere, 106, *125, 132,* 135, 138, *139,* 147, 157, *158,* 164, 168
Atlantic coast, 17, *25,* 29, 92
Atlantic Ocean, 19-20, 22, 27, 45, 75, 82, 87, 92, 96-97, 120, *122, 125,* 133, 141, 146, 174; floor, 141, 169; North, 70, 169
Atlantic valley, 19
Atlas Mountains, *25*
Australia, 19, *26,* 39, 41, 46, 70, *84,* 124, 131, *143,* 143, 145, *152-153*
Australian plate, 131
axis of rotation, of earth, 71-72, 74-75; displacement of, 73
Azores Islands, 120, 146, 169

Bacon, Francis, 17-19, 29, 37, 121, 175
Baja (Lower) California, *27,* 82, 116
Baker, Howard B., 24, 28
basalts, basaltic rock, 77, 109, 124, 126, 135, 159-160, 162
Becquerel, Henri, 102
Belusov, V., 109
Benson, William E., 177-178
Bering Sea, *26*
Bismarck Archipelago, 130
Blackett, P. M. S., 109
blocks. *See* tectonic plates
Bouvet Island, 120, 169
Brazil, 18-19, 68; coast of, 37
British Isles, 17, 116, 118
Bullard, E. C., 103
Burma, *26*
Bykhanov, Y., 20

Index

California, 27, 62, 82, 84, 89
Cameroons, 37
Canada, 27, 111, 176
Cape Fortune, 82
Cape Horn, 27
Cap São Roque (Brazil), 37
carboniferous era, 72
Caribbean plate, 176
Caribbean Sea, 27; region, 176
Celebes, 26
Cenozoic geological period, 153
Central America, 16, 41-42, 130, 132, 141, 143, 143, 145, 159
centrifugal force on Earth, 21, 74
Chile, 27
China, 136
cities shift or drift, 72
Clegg, J. A., 114
climate, climatology, 36, 70, 72
Clipperton Island, 128
coal, coal beds, 20, 72, 78, 93
coastlines, coasts, 19, 60, 153
Colombia, 130
compression. See tectonic plates
conduction. See heat
Conrad Discontinuity, 52
continental blocks. See continental plates
continental crust, 43-44, 46, 49, 51-52, 55-57, 59-61, 63, 65-66, 73, 75, 77, 82, 86-88, 92-93, 95-97, 99-103, 105-107, 109, 115, 125, 132, 134, 136-138, 139, 146, 148, 157, 158, 159-160, 162-163, 168-169, 180; composition of, 64-65; drift of, 73; measurements of, 55, 56, 60-61; radioactivity of, 101-102, 105; temperatures of, 93, 101, 120, 163
continental drift and drift theory, 18, 21-22, 24, 25, 27, 28, 29-35, 37-41, 43-44, 52, 63, 65, 68-70, 72-73, 75-76, 79, 91-93, 95, 102-103, 108-110, 112, 115-117, 117, 118-120, 148, 150, 154-155, 156, 171, 173, 174, 176, 177, 179-180; rate or speed of, 38

continental plates, 49-50, 65, 74, 79-80, 87, 97, 103, 115, 135, 143, 162. See also tectonic plates
continental rotation, 118
continental shelves, 21, 49, 60, 78, 148, 153
continents, 16, 19-21, 24, 27, 29, 40-41, 43-44, 46-47, 48, 49, 52-53, 60-61, 64, 68, 70, 74-76, 78-79, 84, 93, 97, 106, 109, 118, 120, 125, 129, 136-138, 143, 145, 148, 149-150, 153, 156, 160, 162-164, 170, 180; growth of, 129; "lost" or sunken, 21, 61; motions of, see continental drift; origin of, 30; sites of, 18, 20
convection, convection currents, 75-76, 92-97, 124-125, 128, 131, 132, 135-138, 147, 157, 159, 160-161, 162, 163-165, 167, 167-168, 170; convective plumes or pipes, 165, 167, 168, 170
core of earth, 20, 65-66, 87, 96, 100-101, 127, 158, 160, 162-163, 165, 167; composition of, 65, 66, 160; density of, 65-66; measurement of, 65; temperature of, 100-101
Cretaceous geological period, 39, 71-72, 152
Crete, 146
Croatia, 51
crust of earth, 20, 30, 35, 43-44, 46, 49, 51-52, 55-57, 59-61, 63-64, 134, 161; composition of, 64; crustal matter, 25, 45; crustal plate movements, 37, 40, 134; displacement of, 30; radium content of, 101. See also continental crust; ocean floor crust
Cuba, 42
Curie, Marie and Pierre, 102

Darwin, Charles, 21, 179
Darwin, G. H., 21-22
Deccan plateau, 115
Deffeyes, Kenneth S., 164
diamagnetic, 112. See also magnetism

Index

Dietz, Robert S., 148, *149, 154,* 180
Dominican Republic, 42
drift. *See* continental drift
dunite, 51
du Toit, Alexander, 68

earth, 21, *25,* 44, 63, 65, 67-68, 70-71, 73-75, 78, 84, 86, 88, 91-93, 95-96, 100-101, 103, 106, 109, 111-113, 123, 128-129, *133,* 134-135, 137, *139,* 140, *141,* 145-148, *149-150,* 154-157, *158,* 159-160, 163-165, 168, 171, 177; age of, 22-23, 128, 168, 178-179; core of, *see* core of earth; crust of, *see* crust of earth; evolution of, 66-67; magnetism of, *see* magnetism; mass of, 21; origin of, 43, 65; radius of, 65, 88; rotational axis of, 71-72, 74, 116, 140
earthquakes, earthquake activity, 63, 75, 84, *84,* 86-87, *88,* 88-92, 121, 123, 128-129, *132,* 136, 140, 147, 156; origins of, 86-88, 121
Easter Island, 128
East Pacific Ocean floor, 128
East Pacific plate, *84,* 131, 143
East Pacific Rise (or Ridge), *122,* 128, 144, *159, 176*
El Salvador, *132*
England, 77, 114-115, 174, *174*
Eocene geological period, 71-72
Eötvös, Baron Roland, 75
ESSA (Environmental Sciences Services Administration), *84*
Eurasia, 38, 46, 82, 96, *115,* 116, *151,* 169
Eurasian plate, *84,* 87, 116, 131, 141, *143,* 144-146
Europe, 16-19, 22, *26,* 27, 38, 46, 73, 97, 116, 118, 120, 136, 141, *152,* 169, 173, 180; drift of, 72

faults and fractures, *25,* 78, 81-84, 86, 89, 92, *122,* 124, 128, 138-140, *150;* San Andreas Fault system, 82-83, *83,* 84; San Francisco Earthquake fault, 82; strike-slip faults, 82
ferrimagnetic, 112-113. *See also* magnetism
ferromagnetic, 112-113. *See also* magnetism
Fiji Islands, 130-131
Fisher, Osmond, 21-22
fissures, *133*
folding of mountains. *See* mountains
Folkestone, England, 174
fossils, 20, 36, 70, 174, *174*
fractures. *See* faults
fracture zones, 127
France, 18
Frankfurt am Main, 30

Gambier, French Polynesia, hot spot, 169
gamma, 111
gaps, 81-82, *84,* 164, 173
gauss, 111
geodesy, 35
geological dating, 39
geology, geologists, 35-36, 38-39, 44, 56, 62, *62,* 63, 66, 68-69, 71, 81, 97, 102, 108-109, *115,* 124, 129, 131, 134, 137, 140, 156, 171, 173-175, *176,* 178-180
geophysics, geophysicists, 19, 28-29, 33, 35-36, 38, 47, 50, 53, 56, 60, 65, 69, 75, 86-87, 93, 97, 100, 103, 105-106, 108-109, 115, *117,* 121, 128, 131, 134-135, 156, 162, 171, 173-175, *176,* 179; measurements, 144, 145
geothermal gradients, regions of, 100
geysers, 87, 156
Gibraltar, 146
Glacier Peak, *63*
glaciers, 78
global tectonics. *See* tectonics
Gondwanaland, or Gondwana, 39, 97, *150,* 160, *161*
Gough Island, 120
Graham, David, *176*

Grand Banks, 173-174, *174*
Grand Canyon, 78
granites, 77-78, 109, 137, 159-160, *161, 162*
gravitation, gravity, 106-107, 120, 129, 165, 179; acceleration measurements, 106; attraction, 107, of star, 21; energy, 160; force, 107, 120, 129; shortages or deficiency, 106-107
Greece, *25*
Green, W. L., 20
Greenland, 27-28, 31, 33, 38, 49, 76, *152*
Guam, *26*
Guam Trench, 107
Guatemala, *27*
Gulf of Aden, 121
Gulf of Antilles, 19
Gulf of California, *27*, 81-82, *122*, 128, *159*
Gulf of Guinea, 18-19
Gulf of Honduras, *27*
Gulf of Venezuela, *27*

Haiti, 42
Hamburg, Germany, 87, *88*
Hawaii, hot spot, 169
Hawaii Trench, 107
heat: conduction and conductivity, 92, 95, 98, *99*, 99-100, 127, 135, 163-164; convection currents, *see* convection; energy, 100, 103, 147, 155; flow, *104*, 105, 107, 120-121, 129, 135-136, 147, 155-157, 159-160, 162-165, 169; gradients, 101; power, 104, 157; radiation, 155; units of, 11-12. *See also* temperature; thermal activity
Hess, H. H., 154
Himalaya Mountains, *25*, 40, 47, 69, *115, 143*, 146, *153*
Hokkaido Island, Japan, 130
Holden, John C., 148, *149*
Honduras, *132*
Honshu Island, Japan, *26*, 130

hot spots (convective plumes or pipes), 107, 164-165, *167*, 168-170
hot springs, 87, 156
Humboldt, Alexander von, 19
Humboldt Current, 19
Hutton, James, 178
hydrosphere, 53
hypsometric curve, 47; graph, *48*

ice crust, 33
Iceland, 120-121, 169, *176*
India, *26*, 39-40, 69-70, *84*, 115, *115*, 116-117, 138, 141, 144, 146, *150-153, 158*
Indian-Australian plate, 144-146
Indian Ocean, 45-46, 121, 124, *158-159*
Indian Ocean ridge, 144
Indian plate, 141, *143*, 143-144
Indochina, 86, *143*
Indonesia, 69, *159*
inland seas, spreading and dwindling of, 92
International System of Units (SI), 11, 111
Iran, *25*
Iraq, *25*
Ireland, 38
islands, 19, 24, *26-27*, 41-42, 46, 64, 92-93, 107, 118, 120-121, 143, 145; origins of, 41
isostasy, 56-59, 106-107, 129
isotopes, 103-104
Italy, *25*
Iwo Jima, *26*
Izu Trench, 130

Japan, *26*, 115-116, 118
Japan Trench, 130-131
Java, *26*, 131
Java Trench, 131
Jeffers, Robinson, 21, 68, 89-90, 171
Jeffries, H., 108
Jurassic geological period, 39, 72, *151, 173*

Kamchatka Peninsula, 130
Kanamori, H., 109
Kenya, 19
Kermadec Trench, 131
Kirsch, G., 97
Köppen, Professor Wladimir, 29, 31-33
Kunlun Mountains, *115*
Kuril Islands, *26*
Kuril Trench, 130-131, 145

land bridge theory, 29, 70
Laurasia, *150,* 160, *161*
lava, 63, 93, 115, 147, 157, 169
Leipzig, shift of, 72
Le Pichon, Xavier, 134, 140, *141, 143,* 145, *177*
limestone, 77
Link, Peter, *176*
lithosphere, 53, *132,* 138, *139,* 146, *158,* 168-169
London, 112
Los Angeles, 82
Lower (Baja) California, *27,* 116, 137
Lyell, Charles, 178

Macquarie Island, 146
Madagascar, 69-70, 144
magma, 62-63, 78, 81, 92-93, 113, 128, *133. See also* rocks
magnetic field. *See* magnetism
magnetic flux density. *See* magnetism
magnetism and magnetization, 110-119, 126-127; behavior of, 112, 126; directions of, 111, 113-114, 116, 118, 127; of earth, 111-116, 126-127; intensities of, 111-112, 126; magnetic fields, 111-117, 126-127; measurements, 111, 113, 127, 144-145; paleomagnetism, 113-117, *117,* 119; patterns of, 112-113, 116, 126; remanent, 113-115, 118, 126-127; reversal, 126-127; spontaneous, 112; thermoremanence, 113
magnetometers, 112-113, 126-128

mantle, 43-44, 51-53, 55-57, 63-65, 74-77, 79, 81-82, 87, 92-93, 95, 97, 100-103, 105-106, 113, 121, *125,* 128, *132-133,* 134-136, 138, *139,* 145-147, 157, 159-160, *161,* 162-165, *167,* 168, 179-180; asthenosphere of, 106; rock measurements, 56; radioactivity of, 102; temperature of, 101, 106
Mariana Islands, *26, 130*
Mariana Trench, 47, 130-131, 145, *159*
marine geology, 179
mass of earth, 21; measurement of land area, 60
McKenzie, D. P., 134
Mediterranean Sea, *25, 122, 143,* 146
megashears, *150*
Mercator, Gerhard, 17
Mercator projection, *143, 149, 167*
metamorphic rock, 77-78, 113-114
Mexico, *27,* 82, 137, *159*
Mexico City, 130, 145
micropaleontology, 174, *174*
microseisms. *See* earthquakes
Mid-Atlantic Rise (or Ridge), *84,* 121, 123, *123,* 136, 141, 144, *158,* 169, *176,* 180
Middle American Trench, 130
Minas Gerais, Brazil, 68
Miocene geological period, 71-72
Moho Boundary (Mohorovičić Discontinuity), 51-53, 55, 64-65, 79, 103, 105, 146-147, 163
Mohorovičić, Andrija, 51
moon, 21-22; earth birth theory of, 21-22; mass of, 21
Morgan, W. Jason, 134, 164-165; Morgan model, 164-165, 168
Mount Adams, *63*
mountain(s), *25,* 27, 40-41, 52-53, 55, 58-59, 68, 76, 129, 156; chains, systems and submarine, *25-27,* 40, 79, 121, 123; floating of, 58-59; folding of, 79-81, 91, 109, *115;* making or form-

188 Index

ing of, *24-25,* 30, 41, 76, 91, *115,* 146; movements of, 24, 30
Mount Baker, *63*
Mount Hood, *63*
Mount Lassen, *62-63*
Mount Ranier, *63*
Mount St. Helena, *63*
Myeda, S., 109

Naja Hills, Burma, *26*
New Britain Trench, 130
Newfoundland, 38, 173, *174*
New Guinea, *26,* 41, 130, 146
New Hebrides Trench, 130
New World, 16, 22, 145
New York City, 111
New Zealand, 41, 70, 131, 143, 145
Nicaragua, *132*
North America, 16-19, 27, 38, 40-41, 46, *62,* 75, 82, 97, 117-118, *132,* 136, 141, *143,* 151-152, 169, 173, *176*
North American plate, *177*
North Atlantic Ocean, 70, 136, *152,* 169; floor, 128, 169
North Atlantic Ridge (or Rise), 128
Northern Hemisphere, 75, 172
North Indian Ridge (or Rise), 128
North Pacific Ridge (or Rise), 128
North Pole, 70, 72, 111, 140
North Sea, 87
North Tonga Trench, 146
Novum Organum, 17

Obruchev, V., 109
ocean floor, 21, *25,* 29, 42-43, 46-48, *48,* 49-50, 52-53, 60-61, 63, 79, 87, 97, 107, 109, 120-121, 123-124, *125,* 125-129, 131, 134-137, 141, 143-145, 148, 155, 162-163, 170, *176,* 179; measurement of, 60, 120, 127-128; spreading of, 144, *153-154,* 156, 179
ocean floor crust, 43-45, 49, *49,* 51-53, 55-56, 60-66, 74, 77, *84,* 87, *88,* 92-93, 96, 100, 105-107, 113, *115,* 120-121, 124, *125,* 126-129, 131, *132-133,* 133, 135-138, 145-146, *150,* 157, *159,* 159-160, *162,* 163-164, 168-169, 179-180; age of, 128-129; convection of, 124; measurements, 55, 56, 60; radioactivity of, 105; temperatures of, 93, 120, 163
ocean floor plates, 79, 124, 135
oceanic rises. *See* rises
oceanography, oceanographers, 45, 120, 172
oceans, 21, *26,* 29, 39, 43-46, 53, 60, 64, 73, 121, 126, 129, 131, 133, 136-137, 155-157, 160, 163; building of, 169; density of, 45; mass of, 45; volume of, 61
ocean trenches, *84,* 87, 92, 107, 121, *122,* 129-130
ocean water, measurements of, 56
Oregon, *27, 63*
Origin of Continents and Oceans, The, 32, 35
Ortelius, Abraham, 17

Pacific-Antarctic Ridge (or Rise), 128
Pacific area, 87, 92; central Pacific, 116
Pacific coast, *27, 62,* 86, 92, 121, 130, 136, *143*
Pacific Ocean, 19, 21-22, *27,* 40-41, 45, *62,* 82, *84,* 86-87, *88,* 92, 107, 121, 124, 129, 131, 133, 143-146, *153, 158;* floor of, 21-22, 41-42, 47, *159,* 172-173
Pacific Ocean plate, *176*
Pacific plate, 84, 131, 136, 141, *143,* 143-145, 169
paleoclimatology, 32, 35-36, 69
paleomagnetism, paleomagnetists, 113-117, *117,* 119, 127, 179-180. *See also* magnetism
paleontology, 35-36, 69, 177, 179
Palmer Peninsula, *27*
Panama, 130
Pangaea, *149-150*

Index

Panthalasia, or Panthalassa, *162*
paramagnetic, 112. *See also* magnetism
Parker, R. L., 134
Pelvoux Massif, 55
Permian geological period, 72, *149*
Peru, 17, *27*
Peru-Chili Trench, 130
Peru Trench, *159*
Peterson, Melvin, 177
Philippines, *26,* 130, *159*
Philippine Trench, 107, 130, *159*
photosynthesis, 156
physical geography, 19, 45
phytogeography, 35, 37, 177
Pickering, William H., 22
Placet, François, 18-19
plates. *See* continental plates; ocean floor plates; tectonic plates
plate tectonics. *See* tectonics
plumes or pipes. *See* hot spots
Point Gallinas, *27*
polar wandering, 70, 72, 117-118
Portugal, 173
Powell, D. E., 173
Precambrian geological period, 116
Puerto Rico, 42

Quaternary geological period, 71-72

radioactivity, 97, 101-103, *104,* 105, 135, 147, 157, 159-160, 162-163; measurement and dating of, 38, 102, 104-105, 115, 124, 127
radium and radioactivity, 101-105
rafts. *See* tectonic plates
Red Sea, 81-82, 137, *153*
ridges, ridge systems. *See* rises
rifts, rifting, rift systems, 81-82, *84,* 86, 121, *123,* 124-125, 127, 131, 137, *152-153, 158-159,* 164, 174, 180
"ring of fire," *62, 84,* 86, *88,* 121, 129
rises (ridges), *25,* 121, *122,* 123, *123,* 124, *125,* 125-129, 131, *133,* 134-136, 140, *143, 150, 158,* 160, *162,* 164-165, *167, 176,* 179; measurement of, 123, 128; ridge-rift systems, *152-153, 159;* ridge systems, 128, 133, 137-138
rock dating, 37-38
rocks, 50, 66, 68, 73, 75, 77, 79, 84, 112-113, 115-116, 118, 124, 126, *139,* 172, *174;* age of, 66, 126, 129; basaltic, 77, 105, 109, 115, 124, 126, 160, 162; crustal, 124, 138; granitic, 77-78, 105, 109, 160, *161,* 162; igneous, 62, 78-79, 113-114; limestone, 77; magmatic (magma), 62-63, 78, 81, 92, 113; magnetism of, 110, 112; mantle, 65, 93; metamorphic, 77-78, 113-114; nonigneous, 77, 113; radioactivity of, 102; reverse magnetism of, 126; sandstone, 77-78; sedimentary, 77-78, 113-114, *115,* 117, 124, 160, 172; sialic, 105, *115,* 160, *161;* simatic, 105, 113, *115,* 160; subcrustal, 97; suboceanic, 50, 74; ultrabasic, 74, 105, 157, 160; volcanic, 62
Runcorn, S. K., 116-117, *117*
Rutherford, Ernest, 102
Ryuku Trench, 130

Samoa, 131
San Andreas Fault, 82-83, *83,* 84, 136
San Francisco, 82, *83,* 84
San Francisco Bay, 82
San Francisco Earthquake, 82, 84, 86
Scandinavia, 59
Schwimmer, R., 95
Scotia Trench, *149*
sea-floor spreading. *See* ocean floor spreading
sedimentary rock. *See* rocks
sediments, 59, 60, 77-79, *115,* 124, 129, 138, 160, 172-174, *176;* age of, 129; measurements of, 56
seismic: activity, *63,* 84, 92, 133; records, 87; science, 86; shift, 89; waves, 87

seismology, seismologists, 51, 81, 86, 156, 179
separation of plates. *See* tectonic plates
shears, *25, 150*
SI (International System of Units), 11, 111
sial, 50-51, 60-61, 78, *161;* sialic crust, 55, 60, 63, 77, *115,* 159
Siberia, 116, 130, 145
Sibirga, G. L. Smit, 69
Sicily, *25*
sima, 50-52, 60-61, 75, 135; simatic crust, 55, 60, 63, 65, 74, 77, 113, *115,* 159; simatic layer, 53
Snider-Pellegrini, Antonio, 19
South America, 16-20, *27,* 29, 37, 38, 40-41, 46, 68-71, *71,* 72, 75, 79, 82, *84,* 111, *122, 125,* 128, 130-131, 133, 137, 139, 141, *143,* 143-145, *149, 151, 159,* 169, *176-177*
South American plate, 136, 177
South Atlantic Ocean, 82, 128, 136, 141, *152,* 169; coasts, 22; floor, 128
South Atlantic Ridge (or Rise), 128
Southern Hemisphere, 75, 169, 172
South Georgia Island, *27*
South Indian Ridge (or Rise), 128
South Kermadec Trench, 146
South New Zealand Trench, 146
South Pole, *27,* 41, 46, 70-71, *71,* 72, 140, 143
South Sandwich Islands, *27, 177*
Soviet Union, 32
St. Helena Island, 120, 169
Straits of Magellan, 17
strike-slip faults, 82
Stubbs, P. H. S., 114
subduction, 131, *132,* 139
subocean floors, 51-52, 56
Sumatra, *26,* 131
Switzerland, *25*

Taiwan, *26,* 130
Takeuchi, H., 109, 164

Tanzania, 19
Taylor, F. B., 24, 27-28
tectonic plates, *25, 80,* 80-84, *84,* 86, 124, 131, 134-139, *141,* 141, *143,* 146-147, 157, 162, 169, 180; compression of, 92, 96, *143,* 145-147, *150,* 159; movements of, *25, 26-27,* 40-41, *80,* 80-81, *139,* 140, *143,* 145, 147, 164-165, *167,* 168, 176; rotation of, 144-146, rate of, 144, *176;* separation of, *143,* 144, 146-147, *158-159,* 160, 164-165, *167*
tectonics, 16, 44, 69, 76, 79, 81, 88, 108, 109, *117,* 131, 134, 136, 138, 147-148, 154, 165, 170, 175, *176;* global, 18, 40, 76, 88, 108, *117,* 131, 134, 148, *154,* 156-157, 172-173, 175, 177, 179-180; plate, 18, 134-136, 179-180; tectonicians, 140, 148; total, 134, 138, *139,* 146
temperature, temperature differences, 92-98, *99,* 99-101, 103, *104,* 105-106, 114, 135, *139,* 147, 155, 159-160, 163, 165, 168; gradients, 100
Tertiary geological period, 24
tesla, 111
Tethys Sea, *115, 149*
Texas, 117
thermal activity and conductivity, *63,* 87, 97-98, *99,* 99-101, *133, 151,* 156, 162, 165, 169-170; measurement of, 98. *See also* heat; temperature
thermoremanence, 113. *See also* magnetism
thorium, 102, 104
Tibet, *115,* 146
Tonga Trench, 131
tremors. *See* earthquakes
trenches, deep-ocean, *84,* 87, 92, 107, 121, *122,* 129-131, *132,* 135, 138-140, 145-146, *150, 153, 159, 162,* 164, *176-177,* 180
Triassic geological period, 39, 72, 114, *150,* 173

Tristan da Cunha Island, 120, 169
troughs. *See* trenches
Turkey, 146

ultrabasic rock, of mantle, 105, 157, 160
Unalaska, 118
uniformitarianism, 178-180
uranium, 102-103
U.S.A., 136
U.S. Environmental Services Administration, 148
Ussher, Bishop, 23

Vine, Fred J., 128
Vityaz Trench, 130
volcanoes, volcanic activity, *62, 63,* 75, 87, 92-93, 100, 120, 129, *133,* 147, 157, 169, 173; regions of, 63

Walvis hot spot, *151*

Washington state, *27, 63*
Wegener, Alfred, 28, *28,* 29-4*;,* 43-44, 46-47, 49, *49,* 50-53, 60-61, 65-66, 68-77, 79, *80,* 80-82, *83,* 86-87, 92-93, 95, 97, 101-103, 105-108, 113, *115,* 116, 119, 126, 138, 148, 174-175, *176,* 177
West Indies, *27, 122*
West Java Trench, 146
Wettstein, H., 20-21
Willis, Bailey, 109
Wilson, J. Tuzo, 175, *175*

Yap, *26*
Yap Trench, 107, 130
Yellowstone, Wyoming, 170
Yugoslavia, 51

Zagros Mountains, *25*
zoogeography, 35-36, 69, 177

About the Author

H. ARTHUR KLEIN was born in Manhattan and lived briefly in Nebraska and Europe before his family settled in Southern California. He majored in English at Stanford University and received a Master of Arts degree from Occidental College at Los Angeles.

Mr. Klein worked in London and Berlin as a reporter and feature writer for news services for several years, and then returned to the United States to work as a publicist, writer, and college teacher. He now devotes full time to writing and is an active member of the National Association of Science Writers. Mr. Klein's other books in the Introducing Modern Science series include *Masers and Lasers, Fuel Cells, Bioluminescence,* and *Holography.* He is also the author of *Surfing* and of *Surf-Riding.*

The Kleins live on the beach at Malibu, California.

Dr. Crippen's Diary

Also by Emlyn Williams

Plays

A MURDER HAS BEEN ARRANGED
SPRING, 1600
NIGHT MUST FALL
THE CORN IS GREEN
THE LIGHT OF HEART
A MONTH IN THE COUNTRY (Adaptation)
THE DRUID'S REST
THE WIND OF HEAVEN
TRESPASS
ACCOLADE
SOMEONE WAITING
THE MASTER BUILDER (Adaptation)
CUCKOO

Books

GEORGE, an Early Autobiography
EMLYN, an Early Autobiography
HEADLONG, a novel
BEYOND BELIEF, a Chronicle of Murder and its Detection

Dr. Crippen's Diary

AN INVENTION

Emlyn Williams

Robson Books

To

ALAN WILLIAMS

fellow-diarist
(THE BERIA PAPERS)

First published in Great Britain in 1987 by Robson Books Ltd., Bolsover House, 5-6 Clipstone Street, London W1P 7EB.

Copyright © 1987 Emlyn Williams

British Library Cataloguing in Publication Data

Williams, Emlyn
  Dr. Crippen's diary.
  I. Title
  832'.912 [F]   PR6045.I52

ISBN 0-86051-407-2

All rights reserved. No part of this publication may be reproduced, stored in a retrieval system, or transmitted in any form or by any means, electronic, mechanical, photocopying, recording or otherwise, without prior permission in writing of the publishers.

Typesetting by Bookworm Typesetting, Manchester, England.
Printed in Great Britain by
St Edmundsbury Press Ltd, Bury St Edmunds, Suffolk.
Bound by Dorstel Press Ltd, Harlow, Essex.

Contents

FOREWORD

PART ONE
Towards Scotland Yard

1	THE GUY NOBODY NOTICES	15
2	CURIOSITY	27
3	INDIFFERENCE	38
4	LOVE	48
5	HATE	60
6	I'VE ALWAYS BEEN TIDY	69
7	OVER AND DONE WITH	81

PART TWO
The Yard Versus Crippen

8	JUST ROUTINE	105
9	YARD NOT SO FRIENDLY	125
10	THE NEW LIFE	152

Afterword

1 THE REST OF THE FACTS 167
2 FACT AND FICTION IN
   THE DIARY 172
3 THE WOMAN IN THE CASE 176

Foreword

BEFORE TURNING TO the next pages, the reader has the right to ask a question. 'What is this? Here's one of the best-known murderers in history – is his "Diary" a genuine document?'

My answer, at the moment, is 'Yes, it is'. As is indicated towards its end, the Diary was to remain hidden until the year 1985, when it would come to light.

And has. Doctor Crippen's idiosyncrasies, in the way of punctuation and spelling, have been retained – the voice is his, speaking on paper, and should be heard without interference. It is his Diary. Luckily, those idiosyncrasies are not extreme enough, or recurring enough, to become an irritant.

That's my story, and I'm sticking to it. I hope the reader will too.

But the reader, having done just that and come to the end of the Diary, will be further entitled to ask a couple of detailed questions about it. These I propose to answer in an 'Afterword'. Till then—

Emlyn Williams

Thank You

Alex Beauclerc, Ursula Bloom, Peter Bull, Larry Bussard, Robert Crawley, John deLannoy, Michael Down, Nigel Norland, Frederick Nicolle, Ann Plugge.

PART ONE

Towards Scotland Yard

1

THE GUY NOBODY NOTICES

Hospital College, Cleveland, Ohio

March 13, 1883

This is me, Hawley Harvey Crippen & what do you know, I am 21 today! (The 13th, but Im not superstitious.) And since there is nobody in this sleazy rooming-house to tell the big news to, heres me telling it to myself.

And in a Diary!! What am *I* doing with a Diary? Because its my only Birthday Present, thats why. When you are on your ownsome & dont mix, even your 21st B. can pass unnoticed.

Except by Mr. Melton my English teacher in high school – he has kept up with me ever since, with birthday cards & now this, he's a kind fanciful soul. His note says, Make sure you keep this Diary. First put your particulars, like filling up a form for the authorities, then various happenings & your thoughts about same, you never know where it might lead.

The reason old M. got the idea I could be a sort of writer in this way was that he once gave us a Composition to do, the *Adventures of a 5-Cent Piece*. Well, I was only 12 but I took that Coin all over town – out of the cash-register into the purse of a rich lady covered in jewelry, then its stolen by 2 burglars who fight over it & one kills the other & old M. said that was far-fetched & better stick to facts.

But he did say, in his blunt way, Crippen youre nothing to look at & not blessed with a pretty name & never open your mouth, but stick a pen in yr hand & youre off. So here I am, back to the same ruled paper. Try anything once.

(Mind you, a couple of 100 yrs ago my name was *Crispin*, & thats a goodlooking name. Too bad that extra *P* creeping in.)

I do know one thing this writing will show up – faulty punctuation & certain spellings not what they might be. Funny if I turned out a Famous Diarist like that *Rousseau* with his Confessions – except he put in dirty things & that I dont hold with & never will.

What shall I put? Born Coldwater Michigan, father Myron Augustus Crippen (Myron, good old Amer. name!) dry-goods merchant. In *Cold*water – dry & cold, thats me. Not a v. good start in Life. Dont remember much of my folks, mother died & Dad moved to L.A. & we lost touch. Ive always been pretty much on my own. (Now thats pretty darn dull, how am I to keep this up?)

But Ive got an idea that at this moment, just putting the words together is helping to simmer my temper. Because I'm pretty mad, my hand is shaking. Just now, stepping out of the elevator, I overheard a drunk say, That little runt got a moustache that would look better on a walrus. I despise him, & me too for this shaky hand. I'll raise that hand now till it obeys me & steadies up.

Well, Ive got to the bottom of that ruled first page & starting another, what do you know? And I feel better.

Mr. M. was right, Ive never talked & never will. I hear all the chitchat around me & am amazed at the wasted energy, like Niagara – I mean, whats the point? I listen & I criticize, in my opinion, it's more restful. My hobby as you might say.

It's funny that while I size everybody up around me, I've been called the Guy that Nobody Notices.

And that suits me, I know my place. It even shows in what I think of my 1st names. For reasons best known to his own self, my father had me christened Hawley Harvey. Sounds like a dude actor, & when people rudely ask me what H. H. stands for, I always say, Henry Harold.

Thats more down my street, ordinary. Folk dont look straight through me, they just bend to left or right to see who's behind me, like I was a lamp-post. Not that tall a

1883: CLEVELAND, OHIO

lamp-post neither, perhaps that's why I keep to myself. If I was 6 foot in my stockinged feet, would I be maybe sitting at the bar playing poker with the Big Boys?

Silly question, because here I am, 5ft 4½ (just), bulgy forehead, weakish moustache sandy-colored (that same Walrus Attachment) Oh, I forgot the main point – thick moony steel-rimmed glasses. The other day some fresh guy said they're so thick you cant tell the color of my eyes, so when I get back to my room I take them off to find out, & of course I *cant* find out because I cant see, I guess they must be sort of watery grey. Folks also tease me about my walk, they say I walk with feet pointing out, which must amuse people behind me. What's more, I'm losing my hair (on my 21st birthday!)

Oh I *have* a kind of sense of humor, if folks only knew it – but they cant know it, because I dont talk. (If everybody with a sense of humor was like me, the world would be a quiet place.)

But theres more going on under that hair than meets the eye. For one thing, I face facts – how many men aged 21 will own up that theyre never going to make their mark in the world? To do that, you've *got to talk*, so I'm out right away.

But my intelligence is average, I worked hard at my medical studies & got my M.D. here in Cleveland. Off to London (England) any minute, to perfect them (expenses paid by Uncle Otto, with access to Hospitals so I can watch operations etc, I'm looking forward to it).

Funny how I came to see myself as Doctor – when I was 12 in Coldwater, Uncle Otto was doing well with his barbers shop & I liked to watch the shaving & haircutting. I thought, I'll be a Barber when I grow up, & he gave me lessons in Haircutting after hours (& if I had friends at this moment, I could cut hair for them).

Then comes the day when one of his assistants is drunk & slips while shaving a customer with cut-throat razor & the razor lives up to its name, blood all over the place, my Uncle faints & I rush up with towels & basin & bind up the poor guys neck, tight, they said afterwards I had saved his life.

My Uncle said how brave I was to face the blood, I said it

was just like any other mess, he said I was meant to be a Doctor & here I am, getting ready to be one.

Only not on a big scale, my plan is to be an eye-&-ear man, with Dentistry thrown in, as a boy I used to fiddle with gadgets & things & enjoy getting them to go right, like a Singer sewing-machine one time. And I'm on my way.

Some dope asked whats so great about spying into other people's ears & mouths thats not all they should be, and – as usual – I just didnt say anything. But it got me thinking.

I just dont see those three holes (two ears, one mouth) as any less *objects* than, say, the ins-and-outs in the Sewing-machine. Indeed, when I'm at work, I get quite a surprise when the patient coughs. Like a machine coughing.

Not that I dont know what it feels like, when my own teeth causes me trouble – Ive already got four false ones – but under another dentist I dont budge, pain is something to be wrestled with, like some people fight wild animals. I dont look tough, but I am.

My, my wrist is tired already – thats the beginning, and end, of my diary! What that guy Rousseau must have felt like at the end of them Confessions – But from hints he gives, his wrist got pretty tired other ways, but thats sly talk & only for this ruled paper.

Which I'll be burning anyways. But it has helped my temper.

17 Handel St., Bloomsbury, London
May 12, 1883

What do you know, I somehow didnt burn it & here I am, eight weeks later, having just found it bottom of baggage!

So here I go again.

I have a poor room not too spick-&-span, but I do like this City. After those Middle-West shanty-towns & that terrible twang – I seem to be sensitive to sounds, specially ugly voices – I'm American too but the rare times I open my mouth I am careful to speak quietly & correctly. Here there are others like me, even the Cockneys sound better than at home, & theres parks & squares & trees & nice old horse-busses, & everybody polite.

Except for the woman just now. I guess its temper again this time thats making me race my pen along the lines.

An hour back, I went from my operation lecture into a *pub* for what they call a stout – no hard stuff for me, upsets my stomach – & there was a lady standing next to me.

Now I hadnt gone in there for that, but there she was, a big buxom painted thing with big fake jewelry, drinking a port & lemon. I dont deny I looked her over, I dont approve of her sort but you have to take your hat off to them for being so brazen.

Well, she saw me looking & she said, Youll know me next time, you just take a tip & set your cap at a girl your own bleeding size. I walked out without finishing my stout, I can still hear that laugh of hers as I banged into the swing doors.

But she's the exception. My landlady said only this morning O Mr Crippen I do like the way you say Good Morning, so quiet & pleasant. That gratified me *no end*, as the Londners say.

Then there's the Police. I have been sometimes amused about so many Yankees coming home & paying compliments to the British Policeman, but they were right. Yesterday, on a sight-seeing jaunt, coming out of Westm. Abbey, I strolled round the corner & said to a Bobby, Excuse me. Yes Sir? he said (I guess he said it because I look quite a bit older than 21, but I liked it).

I said, Could you tell me the way to Cleopatra's Needle? And instead of just pointing in a surly fashion, he took trouble to say, Go across the road, round Scotland Yard there, & along the Embankment.

I said, To us across in the States, a yard is the same as what you call a garden here, & I dont see a garden. Then he laughed & said, You wont see Scotland neether, then explained its some sort of Police Headquarters. I was conscious of his courtesy.

I enjoyed the lectures & demonstrations of operations at St Thomas & Barts – one surgeon in particular (French) has taken an interest in me & as well as giving me a couple of side talks on Dissection he talks to me in French & gets me to answer, I'm getting quite gabby at it & he says I have a *flair* for both Surgery & French!! I'm brighter than I thought.

In second-hand shop in Charing-Cross Rd., bought for 6 pence textbook on Dentistry, also (for self improvement) *Cunningham's Manual of Practical Anatomy*.

Walking through Camden Town up towards Hampstead, I took a wander through the residential part, nice fresh air, lovely solid houses in rows, steps up to the front door, bay windows on the first floor – which they call the ground floor here – also on the next floor (which would be best bedroom) then more on top of that.

Plus a front garden with tree, & down from the garden, steps going (under the house) to what they call the *area*, the kitchen & cellar, for storage. Gentlemens houses.

I expect theres such houses in the States, but I never saw any, & these capture my fancy – theyre all in wide curving streets with names like Place & Close & Crescent. Quiet, respectable, everybody keeping themselves to themselves.

Looking up at them, I thought, I must have some London blood in me, Im going to set my heart (if I make the little way I hope to) on moving into one of these, for good, on a life-long lease.

Just like before, I feel better after writing that. It seems crazy to write to myself, like this, things I know already – but it puts it together for you, like framing a picture. Maybe its a habit I ought to take up.

2347 Madison St., Detroit

March 13, 1885

Well, I didnt. (Didnt take up the habit of a Diary I mean, its been 2 yrs to the day since Mr. M. gave me this one.) But then this last yr Ive been hard at it studying for my Diploma, thats what I'm going to N.Y. for, to take it at Opthalmic Hospital.

June 28

After 4 months I've passed! And I'm *Doctor Crippen*, for life!

Passed with Credit, & the Credit said, This Candidate gains special mention for hygienic state of hands when at work on sore gums, & for their soft delicate touch so appreciated by a patient. I never thought of that, must be second nature to me. I'm going to keep my hands up to scratch with that Velvet Lavender they advertise in the N.Y. subway.

Yes, on my way, & my future tall house in the curved London street is that one step nearer.

Its a big house (it's true) for one man to have his eye on. Theyre all family houses. For married couples. Doctor & Mrs. Crippen.

Oh yes, that would suit me – little dinner-parties for professional colleagues & their wives, the dinners served by a maid from the area kitchen, & a nice sensible wine, then whist & a little quiet music on the piano.

The only snag, I cant fit *Mrs.* Crippen into the house – what I mean is, I cant for the life of me get a picture of her.

Of course I *am* troubled by from time to time by what you might call the Flesh & the Devil, but I'm pleased to say I have yet to give way. Yet I dont see myself making a wife of that Jezebel sort that insulted me in the pub, & as for the namby-pamby wishy-washy young girls you see in the park, simpering at each other – no thanks.

If I had a sister to house-keep for me – now that would be the ticket! Doctor & Miss Crippen.

Salt Lake City
March 13, 1888

My 26th birthday, 3 yrs since I wrote the above, & with Detroit behind me (Asst. to Dr. Porter) here I am, married with baby! Shows you dont know yourself as well as you thought you did. Whats the story? Oh, it just happened.

Charlotte Bell. No Jezebel, no Soppy Sal neether – nice respectable daughter of a water-works engineer – well, between me & me & this ruled paper – a Plumber.

She is not pretty & not plain. More plain than pretty – poor complexion – but not a sketch. Taller than me, but what could I expect in a future wife, outside a circus?

She accepted the idea right away – since she is 29, I have an idea her Ma had frightened her by saying she had one foot on the shelf, & she better hurry up & pick up what she could before the store closed.

(No, before the *shop* shut, I like the English word better.)

Well, Charlotte is a nice girl & respects my wishes, & keeps the house clean & cooks alright. I dont love her – at least I dont think so, I wouldnt know as I cant imagine *loving* anybody: *liking* some, yes.

But even then, not enough to make a friend. I cant understand men that are great pals with other men & tell their family headaches & get mixed up in each others lives. Perhaps that makes me sound peculiar, but then it's the Bosom Pals Club that sounds peculiar to me! And which is right?

No, I certainly dont love my wife, going by the way books & soppy poetry go on about *love*. The bedroom business I manage alright, but it doesnt bother me. From the way I overhear other young fellows talk, I must be pretty lenient in that area. And I'm glad, saves a lot of wear & tear. And dough. Enough of that.

And like a lot of married men who dont drool over kids, I was presented with one within a year. A boy, Otto Hawley Crippen if you please, Charlotte wanted him half-named after me & I let it go.

Its hard work here, being an eye-and-ear specialist. No social life – I may not want close friends but I would like to rub shoulders with acquaintances, socially. But the old story – youve got to have the money. I'm not doing bady, but takes time.

When I say Charlotte runs my home, it doesnt take much

for that – a four-roomed dollhouse in a street as straight as a ramrod, & just about as interesting. Whatever a ramrod is.
I still think of London, & that curved road with the trees. Itll come.

Feb. 21, 1891

Charlotte passed away 5 days ago, while I was out of town on business trip (Pneumonia, sick 2 weeks).
That sounds as if I'm writing this (after nothing for nearly 3 years) because I'm upset. We *were* married 4 years. But its not so, I'm writing to put on record – between me & myself as usual! – a funny thing.
Its rare for a doctor to be able to say he's never seen a dead person, but people dont die of ear-ache or a wisdom tooth. So this was my first time.
I walked in & found 3 neighbors (women of course) who had come in to view. They hadnt known her except to nod to, but they all 3 had the same look – a mixed-up thing of mourning whipped up for the occasion – & I guess what they call in books *awe* – ie, fear of a Dead Person.
And they feel this *awe*, because theyve been told to, since they were born. And theres no sense in that fear – what is so mysterious about stopping breathing, when you think it's one of the few prospects weve all of us got in common?
The mortician had done an A.1 job, & I could hardly credit it was Charlotte. She looked far cuter than she ever had in life. More repose. None of those second-rate thoughts flitting through her head – in fact it looked better now it was empty.
He had worked on that poor complexion & she looked downright pretty, like a waxwork – so like, in fact, that I had to pinch one cheek to make sure it wasnt. (The 3 women thought I was giving her a blessing or something – just as well, & one of them started to cry.)
Call me a cold fish if you like, whats the use of me pretending, when I'm telling something *to myself*? It come to me today that not only the mouth & ears, but the whole

human body is just an object, like a clock thats stopped ticking & can never be started again. Doesnt that make sense?

I never took any objection to Charlotte, but she didn't mean much to me. Nor does the boy, he's just a babbling baby. People would say back to me (if I passed such a remark to them, which I would not) Wait now, wasnt *you* once a babbling baby yourself? Of course I was, I got vocal cords, but whats that got to do with it?

Dont mistake me – I'll see he's well cared for, I'm the reason he's here & I'll take the responsibillity, as a moral person.

P.S. A telegram just got to me, from Charlotte's father, saying theyll see to little Otto. Well, I dont like them taking it for granted I cant see to little O. myself. But its a relief.

The funeral was frugal. But rubbish was spoken, from up in the pulpit & by the graveside. They kept referring to *Poor Mrs Crippen*, as if I hadnt ever given her a cent for housekeeping. Charlotte was not afluent, but (thanks to me) not *poor*.

I'm selling this no-good dollhouse, & on to pastures new as they say, St. Louis this time, where I'll take a room. I'm the Wandering Jew alright, anyway the Wandering part.

I miss Charlottes quick step in & out of the kitchen, but living alone will have its points.

St. Louis
10 April, 1891

Well here I am, & doing well on my own, and her gone only 2 months.

And living alone *does* have points. For one thing you improve your mind by reading more – Walter Scott, Pickwick Papers, I like a good yarn.

I also bought a cheap copy of *Dr. Jekyll & Mr. Hyde*. I like a good shocker as well, they take you out of yourself – though this one is pretty farfetched, with one person taking turns being two people, but holds the interest. And a Doctor too, I take that as a libel on my profession! And theres one advantage of living alone, that I never thought of (it's O.K. to mention it here, but not elsewhere). Day or night, you can break wind. I dont mean to step out of place, but it does prove I mind my manners in company.

Hickory Hotel, Brooklyn
March 13, 1892

Here I am again! (I can understand people who keep track of themselves by writing everything down, starting off with Dear Diary, I nearly did it then) Working as dentist for Dr. Jeffrey, & have been for 11 months, doing the best yet – quite a practice, & me well-liked for quietness, courtesy & eficiency.

Im 30 today! I dont mind Middle Age getting nearer, it will suit me better than being young, I think. And I'm all for decency – decent speech, decent deportment, take things & people as they come, no tempers. Thats why I go to the Methodist Church next to my rooming-house – I dont listen to the goody talk but church-going is a respectable habit so long as I dont mix, its a shabby congregation.

As I say, here I am again, pen in hand, but I dont know why, really, since nothing much to report. Just a kind of string of strange faces, mostly in a chair with mouths wide open, I like them best that way because it means that (a) they cant talk, & (b) its money in the bank!

The ones walking about seem to talk more than ever, it beats me.

Getting restless again. Have got the idea of moving on to N.Y. & trying my luck there.

Oh one thing, I'm branching out professionally – I reckon its to do with having found out (in the matter of my wife passing away) that – by nature – I have the sensible scientific attitude towards dead bodies (such as I felt watching those

Ops in London when a student) & therefore an aptitude for Surgery. I'm told a good surgeon can really strike it rich & I'm brushing up on the book I bought in London – Cunningham's Manual.

With luck it might carry me quite a way.

Getting a bit sick of restaurants & sour-faced waitresses & having to wash my clothes at the sink top of the stairs. Might be a good idea to get married again, but the way I live it would be looking for a needle in a haystack called Brooklyn.

Wont write again unless I can put down the name of a new wife.

That makes me smile over my pen, sounding as it does like that British King Henry the Something, that made a hobby of beheading them, & then on to the next – hardly my line of country!

But as I say, I'll wait till I can write down the name of a new wife. A new Charlotte.

April 14

Which has taken me just one month.
Cora Turner!
Isnt that a good-sounding name?

2
CURIOSITY

NOT A WIFE of course – but not another Charlotte neither – Mercy no!

I wish she *was* another Charlotte – & yet I dont – that sounds confused, because I *am* confused, & writing in this may help to straighten me out.

I met her 3 nights ago, when I had been – let me see – 7 months in Brooklyn.

This is the story.

In the week, I sometimes drop in at the Eldorado Vaudeville Theater on the corner – quite a fleabag really, upstairs only a dime. The comics are pretty low but you see the ocasional good act. Seats not comfortable & not much room, last time the guy next to me was squirming and folding himself up – he got quite mad & left.

I suffer no discomfort up there myself, so there are times when it feels good to be below par in the way of inches. Count your blessings.

Well, I was sitting in the Eldorado, up in the dark in the cheap seats, when (for the final act) there comes on a couple called the *Exotic De Sousas* or something, they do a saucy song about Kissing in the Moonlight, then a snaky kind of dance, then he kisses her hand & they sit on a sofa & chat about spooning, then back to the dancing.

Quite daring. But smart, him in white tie & her with an osprey feather or something sticking up out of her head.

The feather should have been across her chest, she was in such a low-cut gown you could nearly see the parting of the ways. If you like that sort of thing, I reckon it could be of interest.

Now I make a point (in such public places) not to look who is seated next to me – but all of a sudden I got a whiff of

strong perfume, like lying on a bed of roses, & had to look.
   I wasnt familiar with perfume, I had disencouraged it with Charlotte, & here was this woman next to me, in the half-dark, quite big, putting away a perfume-bottle & staring at the stage, with her lips not together. Then she caught – I suppose – my thick glasses glinting from the stage lights, turned to me & said, right out as if she knew me – Say, aint that *beautiful*?
   Well, what with that & the perfume I was quite taken by surprise, & before I could think what to say, the Act was over, the couple bowing to the music, then the lights snapped on & everybody getting up to leave. She turned to go and I thought, well thats that, Ships that pass in the Night, when she turns back & smiles straight at me.
   Perfect teeth. (And I ought to know) Then she said, Pea-nut? & offers me one, which I courteously accepted & partook of. She was a good 4 inches taller than me, but now, seeing her in the light, I realised she wasn't an older woman, but a young one, 26 at the most (She was actually 19 – ie, 11 yrs younger than me)
   She had 2 things I was face to face with for the 1st time – & me aged 31! A fur coat, & make-up. Not only that papier-poudry stuff, but rouge & lipstick, the whole kit & caboodle. She looked (at best) like an actress, & at worst – well. But it takes all sorts.
   But her voice was ladylike, when she said, My husband being out of town on business & this neighborhood being a bit scary at night – could I trouble you to walk me to my house on the next street?
   I raised my hat & said, With pleasure, though I was thinking, watch out Crippen, youre no oil painting, she wants to make use of you.
   Robbery? But I only had a couple of dollars on me & my watch being repaired, so I saw her to her house. Glancing sideways at her on the sidewalk, I noticed that she did not pull her veil down over her face, leaving it exposed in the street.
   Then I felt something inside my arm – it was her hand (in a glove) sliding round my arm, as natural as anything. Again

something bran-new for me. A few feet further on, as we strolled along, something new again, & this one took me by surprise, believe me.

A man walking towards us looked at her, stared, & walked past. I turned & looked at her, then back to him, & saw he was looking back as well, at her. That happened twice again, & a 3rd man whistled. She was certainly an eyeful.

And walking along I had to say to myself, look here, whats all this because I realized I was in what the Doctors call a state of arousal – & it wasnt my heart beating either. Unusual for me, take my word. And it was from a woman walking on my arm who was arousing other men.

Her house wasnt hers, of course – a rooming-house just like round the corner. She asked me in, then when she saw me hesitate she said, Oh, only into the parlor.

So I went in. The parlor had a couple of cane chairs & dusty old palms, but she had enough class to make it seem as if she didn't belong.

She threw back her fur & I saw she had a full bust, a bit too full for me, & nearly as low-cut as the actress on the stage. She was wearing a reddish sort of brooch, I asked what it was, she said, Only a garnet but I do love jewelry. Then she brought two small glasses out of a sideboard & poured, it was cordial. Her eyes very dark & soft, like molasses.

Then she talked, but it was interesting, in that quiet voice. Daughter of a Professor in an upstate College, good Catholic family, both parents dead, and she carrying out her Dads ambition for her to be a singer – that was why she had struck me as like an actress – & her husband working hard as the Head of a Travel Agency to provide her with the essential training.

Then I asked her about her voice & she went straight to the upright piano & played & sang. The instrument was out of tune, but the *Rose of Tralee* rang out sweet & clear. As I have remarked, I am a victim to sound, both ways – nice & nasty – and I felt a satisfaction in that musty room.

Then she talked about the Act we had seen, how she

hadnt liked the womans voice, not quite on the note, then she said the reason she doesnt mind going alone like that, is that she is crazy to go on the Stage, and therefore wants to learn from watching the Acts. I said, Arent you afraid of men trying to approach you? She threw her head back with (what they call in story-books) a throaty laugh, & said, Dont worry, I give them the cold shoulder. I thought, you dont look like your shoulders could be anything like cold, neether of them.

Ive never been within a mile of a female like this one. We're meeting next week. I'm puzzled what she sees in me. We'll see.

May 22

After that first night we have met once a week. Its 5 weeks now. Before proceeding to the Eldorado, we met in neighborhood ice-cream parlor (I paid) & after the show the cordial back at her house. The 3rd time she told me she expected her husband back next week.

And the 4th time, she confessed she didnt have a husband at all, & that the man she was expecting back in town (by the name of Lincoln) had taken advantage of her 2 yrs back, when she was 16. He had bought her the fur coat & the garnet, but after that gave her only just enough to live on. He'd promised to pay for her voice lessons, but he was a rat, and she was through with men.

Then we had 2 more cordials, walked over to my (Hickory) Hotel and went quietly up to my room, where intimacy took place.

(What is good about a Diary, you can put in things like that, which you would not dream of uttering out loud.)

It wasnt anything to write home about. I was scared I might not be up to it (if I may so put the problem) & of course that means you havent much of a chance.

But good enough for her to give me a nice kiss, then squeeze my hand. She did say I looked funny without my glasses, I had hoped she would pay me a compliment.

And she did giggle at me keeping my pajama-top on – I reckon it was a bit of a quirk on my part, I just didnt like being in my birthday suit as you might say, we all have quirks.

Then I told her I had good prospects, & proposed. She said Yes. We are to be married in 3 months, Sept 1st.

Its confusing, because I still cant see what she sees in me, though I think she thinks I'm cleverer than I am. I seem to amuse her in some way, Like I was some weird young kid. Youd never think I was 11 yrs older than her.

And what do *I* see in *her*? I certainly dont love her, not the way they carry on in the story-books. I guess that (on my part) its just Curiosity.

I can sort of figure it out though, thinking like this on paper (writing things down things as they come into my head. It feels good as an outlet.)

I want to go on like I was doing that first night when I walked her home. I want to show her off.

1754 Burgoyne St., St. Louis
March 13, 1894

My 32nd birthday.

After we got married (in Jersey City) we came down here, where I been doing well – its 20 months by now, goodo – as consulting physician to an Optician, Mr. Hirsch on Olive St. Was able to give Cora a Xmas present. A diamond ring. Small, but good.

We have a house – the same house that seems to have traveled with us, but weve got to be patient – with some nice new chairs, cheap but goodlooking & over the fireplace a framed painting of George Washington also a photo of Regents Park London.

And Cora keeps house & cooks, not quite as well as Charlotte – but with more life to her, & her singing at her work is really pretty.

And will get prettier I hope, with the voice lessons I'm paying for. I would certainly enjoy sitting in the dressing-room of an Opera Singer, watching folks kiss her hand & feeling that I am the quiet little husband who is back of it all. Nothing wrong with dreaming.

In our private relations, I'm afraid she gets kind of carried away, so I guess its only fair to her that I should perform a husbands duty once a week.

Also, again with her youth & temperament, I can see she craves for company & so I have cultivated 2 or 3 married couples from the local church (Catholic). I'm not religious but its a stylish atmosphere & I go as a social duty. She wants me converted but I've ducked that – meeting it halfway by going along with her to Evening Mass though not to Morning ditto.

None of the friends she has made could be called a Ball of Fire, but civil enough and better than nothing. It means added expense to entertain them, but they ask us back & that equals up.

One of them fancies himself as an Artist, so we have in our Parlor a framed painting of her. She's a fine-looking woman but it gives her a real old simper & all in flounces & a big hat like a Southern Beauty but it brightens up the room & makes a change from George Washington.

Its a dull day-to-day life, but I like that. No more pen-to-paper till further notice.

Feb. 10, 1896

Well, heres the Further Notice, 2 yrs on – no, its not the 10th its the 11th, one a.m., & my hands shaking again.

Its taken all this time to get them shaking, but it had to come. All these phenomenons are gradual, & then – *boom!*

Little things, week to week – little things. Smoking, which I dont hold with in a wife who calls herself a lady. First 3 a day, then 6, then 12, then 20. Me getting back after a hard day to dirty dishes piled up, laundry basket spilling soiled linen & her sitting back with feet up, cigarette glued to her lip & in the sulks.

Drops the voice lessons, the coach isnt good enough, small-town, I'm no hick she says. Gets lonely in the day, wants a cat, I get a cat, she turns against it & the cat knows it & walks out.

Some evenings she beats me at my own game and wont speak! Not a word, not even to answer a question. I try to cheer her up with a glass of her favorite port, no good. The house was small to begin with, but by now its shrinking by the day.

Then she starts going to the toilet, every hour on the hour. I ask if theres something wrong with her bladder, she says, Dont talk to a lady like that, I say I'm not only her Husband but a Doctor, & so we go on.

Then I start noticing her getting v. flushed, & with her putting on weight she looks more like 40 than 23. Then (all of a sudden) she starts jabbering again, & I take note of her voice – that soft purring carry-on in the Eldorado was a performance, & I know too well how ugly noise affects me. The worst Brooklyn twang, like a buzz-saw with a nose.

And with a sprinkle of profanity on top, its not the stuff to soothe you to sleep – Aw for Christ sakes shut your trap! To *me*, the silent one!

Then the chatter – how she misses her father the College Professor, how the only thing she wants is to sing in Opera at the Scala Milan, I am a person whos being starved of what I deserve – only she says *person* as if it was referring to *poison*, stress on the *s – poisson*. And thats not pretty.

Her father the College Professor! One day in the desk I came across her Birth Certificate – he was a college man alright, a Yonkers college where he was janitor. I got from the same Certificate that she isnt Cora Turner at all, she had just picked up that apelation out of a hat. Her real name is *Kunigunde Mackamotzki*, hows that for an unsavory mouthful?

She had to come clean then – her Ma had been German, her Pa a Russian-Polack immigrant. *Crippen* may not sound like doves cooing, but it is a good old English name. *Kunigunde* sounds like something scrawled on the wall of a subway WC.

When I first met her, the only true thing she said was, that she was a Catholic.

Well, on such evenings I used to think (with her gabbling on) – she's hystirical. Then, only last week, I wised up – it wasnt the toilet she was making the trips to, it was the bathroom cabinet. I was looking for some Fruit Salts, when behind the medicines I found a half-empty bottle of Gin.

Well, we come to tonight. The worst yet. I was sitting with the St. Louis paper & having a quiet read, about those terrible goings-on in London a while back – my nice respectable London – with this Jack the Ripper thats a maniac who murdered these women in the East End, terrible to think there are such people about, & they say hes supposed to be a Doctor, like that Dr. Jekyll – but this one is real, I dont like that. A libel on my profession.

Well, I happened to mention this & she says, What sort of women did he murder? And I say (joking), No better than they should be. Then she gives me a glassy look, then snarls through the smoke from that damned cigarette – *Say that again!* I did, adding, You know – prostitutes.

Then she said, going up into a screech – *I know prostitutes?* Are you telling me I'm a poisson that knows *prostitutes?* Then I knew there was no arguing with her.

On and on – jabber jabber about not being 24 yet & her life over before it ever started. She was out to make me mad & I wasnt going to get mad – I'm not the type – & that made her madder. I'm a *lady*, she kept saying, & the more she said she was, the less she sounded like one.

So in the end I said, What *you* need is another date upstairs, with the toilet.

What do you mean, she said, & those eyes that once had looked so melting, now they were like a pair of blackberries on fire – *what do you mean?*

Then I got vulgar back, the way it goes. And I said, When *you* go to the toilet of an evening, its not a case of wanting to get something *out* of your system, its the need of getting something *in*. I think that showed my sense of humor but it did mean the balloon goes up.

She stands in front of the gas fire, sticks her hands on her hips – & they arent hard for her to find neither – and she gives it to me. That black hair was coming down on one side, & she looked a slut. Over her shoulder, the painting of her as a Southern Belle was ogling away at me, & I was interested to compare the two – well, you couldnt.

Then, in the splash of her insults, I thought of her Church friends she was used to billing & cooing with – if they could only see her now! Did I once write down she had class?

Doctor Crippen indeed, she shouts, youre a good-for-nothing small-town quack, youre nothing but a little stick-in-the-mud – & whats more, you've got *me* stuck in the same mud, up to my ass you have, & deeper every day – one thing I hope & pray, Buster – because if not we'll soon be in the gutter – and that is, that youre better at your job than you are between the sheets!

Then she screams – Mr P.T., Mr P.T.! I said, What does *that* mean, & she said, I always call you that to myself, Mr Pajama-Top (referring to my quirk already mentioned, namely keeping it on during intimacy.) Then she said, if a wife isnt ashamed to oblige her husband by showing her tits, what right has he got to insult her?

Then I mention (quite mild, considering her language) that I carry out my duties as a husband once a week regular, which should be enough. Then she said something totally common, so much so I hesitate to commit it to paper.

Once a week, she said, you're damn right about once a week – regular as cockwork!

I was shocked. If *that* is wit, Im glad Im a dull dog. She had moved around the room by now, & this time it was G. Washington looking down at me over her shoulder, cool as a cucumber, as if he was just going to shake his head before saying, *What* a vulgar remark – Crippen, Crippen, what have you saddled yourself with?

She had to stop for breath, so I jumped right in there & asked her why she'd married me. She said, Because she was at her wits end. And then, smart as a whip, she snapped,

Wouldnt I have *had* to be? She told me it had been either having to go be a waitress, or me – so it had to be me.

Then she yelled that she had been scared – dead scared, then she started to blubber, & moan & groan, then I got her to bed & into a stupid sleep. Then me back to front room – I mean parlor, beg pardon Cora. Or should I say Kunigunde?

I get into my chair & I look at old George. And I realized then a funny thing – G. Washington may feel as cool as a cucumber, but so did I.

And I said to myself, Tonight has done the trick. 4 years of a marriage that was like sitting in a miserable little room with the window sealed & the air getting staler by the week – 4 years, & all of a sudden theres an explosion, every pane of glass smashed & a big slab of fresh air, right through.

I sat there as if I was gulping that air in. I thought of the nights I had sat in this chair, mulling over the irritations one by one, & looking at my hands shaking from it. All the little digs & sneers.

No more, everything has come out, I know where I am & what to do.

That sounds as if I was making up my mind to walk out. But people dont act the way theyre expected to, or even they expect *themselves* to. I certainly did not that night.

Because once I'm in a situation, I seem to tend to stay in it. I know I'm moving around the country every couple of years, but thats the circumstances, my ambition is to get set in one place & put down roots & stay.

For instance, though Charlotte was never a big thing in my life, I never thought of my marriage with her ever coming to an end – until it did, the way it did.

And Ive thought just as little of ever trying to end this one – when you think Ive never even stopped to consider what her being a Catholic means; ie that its for life. It just never bothered me.

And it isnt bothering me now. I even feel better than before this blow-up. Hard to say why, but to clear my head I'll try.

When we married I really thought – against those bold looks of hers – that she was a sweet young thing. Then I saw how wrong I was, & over the months got more & more irritated with her. But didnt say anything. Why? Because, inside, I was scared of her. After tonight, no. Because I realized *she's* the scared one. Bored with me – yes, *that* I do know, & maybe cant stand the sight of me – but shes scared. Scared that one of these days she might be left alone. Thats why she hits the booze. Shes dependent. On me.

Its kind of crazy – I can see that to the few acquaintances we have, *she* is the big confident bossy one, with me the timid little husband (I bet the label they stick on me is *henpecked*). Well, its the other way round.

(Right away, Ive got to face one thing – if this drinking business goes on, I'll have to get her into a sanatarium.)

I suppose its like a boxer feels, going into the ring. A challenge.

One thing for sure. Its going to be a long match. The longest ever.

3
INDIFFERENCE

1745 Thomas Ave., Philadelphia
Dec. 31, 1899

HELLO AGAIN, DEAR Diary, after nearly 4 years.

Tomorrow the 20th Century! That sounds like a funny turn of speech, & the idea of 19-this & 19-that! Well, its a good moment to bring myself up to date.

So hello again, after all this time.

I'm going to be 38, her 27. And looks 35, suits me. And a fine 35 too, men look round at her as much as ever.

One of the reasons for that, is that when I thought out the drinking issue, I was being pessimistic. That night – four long years ago – I went upstairs, took the gin out of the cupboard, went down again & put it in the sideboard.

Next morning, no apologies, but I didn't expect any & would have been put out by them. And the gins been in the sideboard ever since, but (except for the rare let-up, when depressed) not over-indulged in, except to keep her going. Ie, when on show to visitors, or as a guest, so she can go on bubbling & flashing & showing those extra-healthy teeth. She can bubble & flash alright.

Of course I should have moved into the smaller bedroom, there & then – but once I didnt, it was difficult not to let things be. But I never performed the Husbands Duty again – I couldnt to save my life, a man is limited that way. I pretended a fellow-doctor had told me I mustnt for my health.

On her good days I was proud of her, & she knew it. Even talked of going back to the voice teacher, but I took the wind out of her sails by saying she had been right, he *was* second-rate & I'm sending you to the best in town. And did. And Im pleased with the result. I gave her a silver bangle for Xmas. I can afford it since I dont spend hardly a cent on myself.

She's not even restless. And I have an idea why – I'm not going to dig my head into the sand. Ive known her well enough when – God help me – I was supposed to satisfy her, to know that – to put it in a coarse manner – she has to have it.

And whats more – & though I dont want to know the details I can make a good guess – she gets it. One married couple in her Catholic group includes a wellbuilt baritone. He may well sing to the Virgin Mary of a Sunday – solo – but (during the week) isnt above a couple of duets on the side. And no virgin in sight.

She and me, we dont get on – but we dont *not* get on, neither, at the moment. I'm not dissatisfied with the arrangement.

March 13, 1900

My 38th birthday.

The 20th Century (cant get used to saying it) has started something for me! In the course of business, Ive become acquainted with a Britisher, a Dr Newton, over from London for 6 weeks on an Exchange Plan – he complimented me on my efficiency & reliability & suggested writing to Associates over there with the idea of fixing me up with something.

And he has! Ive just had my Cleveland Diploma registered in N.Y., giving me the authority (in England) to practice Physic & Surgery – & I am to be sole manager (in London) of *Munyon's Remedies*, a business advertising Patent Medicines – & the one thing in this Century (so Doctors say) which is going to be bigger than Patent Medicines, is the *Advertisement*. And my office is in Shaftesbury Avenue, the heart of the West End!

I sail next week, she follows in a month or so, to celebrate I'm buying her a gold brooch with diamond in centre.

My parlor photo of Regents Park seems to have freshened up. Heres to the future.

34 Store St., London

March 13, 1901

My 39th birthday.

She took 4 months to get here, but I didnt mind, I had digs at 62 Guilford St., Bloomsbury while looking out for a *flat*, as they call apartments here (first essential, piano) – & I had a feeling she was glad of extra time across the Herring-Pond, to keep in training by fitting in as many Duets as possible with a certain Baritone.

The flat is up at the top of a small house near Tottenham Court Road, with a dairy on to the street – poky, but it's London at last.

I got to face it – the London prospects looked rosier the other side of the Herring-Pond than the real thing does. *Munyon's Remedies* is in Shaftesbury Ave. alright, but in a back room of a shabby house at the wrong end, near Oxford Street, & all I have with me is an office boy and a Secretary not much to brag about in that line – a mousy little soul, looks like a schoolgirl but 17, long straight hair that looks as if it's going to get in her inkwell. Pathetic, makes you feel sorry for her.

But it's London, and Cora likes it as much as I do. I catch her trying out a British accent on her second (quiet) voice, such as saying, I guess I'll have a bawth. It amuses me. And I like showing her the Zoo, & Westminster Abbey, as if they belonged to me. Have taken her to the Opera, Covent Garden, Upper Circle, to see Melba.

Just before the overture, a light was thrown on to the Royal Box, and in stepped a handsome man a few years older than me. The audience went mad, everybody got up including us who didnt know his identity. Then I got it from

the whispering – Caruso, Caruso, Caruso. I thought, oh I'd love to be him, in that box, the centre of attention. Now thats unlike me, because I like to lead my quiet life. But we arent the same all the time.
I could be Melba, she said. That night we had a serious talk.
No not about her & me, Heavens above no – about her professional future. She opened up & come clean – something I had sniffed out myself – that Grand Opera was not for her, she was terrified of them big roles & the responsibility, & (though she'd never said it, & wouldnt now) what had attracted her was those great big baskets of flowers being handed up at the end. Also what she was more leery of than anything (again she never said it, nor me) was the hard work.
No, she'd like to go in for what she called, the *lightah stage*. I knew what had done that – I'd taken her to see Vesta Tilley at the Alahambra, & at the Curtain there were flowers galore handed up – so now Variety (thats what they call Vaudeville here) has to be the ticket. I dont mind. Making the rounds of the agents and meeting stage folk will keep her from getting too disagreeable.
Theres not the rush for Patent Medicines that was promised, but London seems good for her housekeeping too – she seems to want to make a go of it. We go from day to day – no excitements, but plain sailing is better than the ups & downs. I'm a peaceful man.

March 13, 1902

40th birthday, so after a year, a birthday calls for a scribble. Whats happened, worth a report? Next to nothing. Made some nice friends, Dr. & Mrs. J. H. Burroughs. I'm quite a bit balder, & Cora's coal-black hair is quite a bit fairer, to do with a bottle. (Not the gin sort, thats still under control.)
Oh, & speaking of hair, my little typist turned up at the Office for her 18th birthday with her hair up & I didnt

recognize her! (Not much of an item for a Diary – and even if it *was* news, it'd be *stale* news, it was 6 months back!)

The future Music-Hall Star Mrs. Cora Crippen has not made the grade yet. Did a sketch she wrote herself at Balham Music-Hall, but very faint applause, a Flop as they say.

I am amused to look at the programme – *Acting Manager, H. H. Crippen*. Do I *manage* her? Her friends wouldnt think so, but she needs me more than she thinks.

She wasnt helped by the name she billed herself by – Macka Motzki, if you please (made up from Mackamotzki). I might just as well go on the Halls as Crip Pen. (Sang song called *An Unknown Quantity*. Judging by her figure – she's putting on weight – if she goes on like this, *An Unknown Quantity* will just about describe her.)

I feel if she is to get anywhere she must have a name which will look well across a marquee, and have got her to turn into *Miss Belle Elmore*. Hello Belle, So-long Kunigunde.

To be honest, she not only has *not* got one foot on the first rung of the ladder, there doesnt seem to be any ladder.

She goes to what they call Auditions (with a couple of gins under her belt) but the pick-me-ups dont seem to do the trick, she comes back with her mouth down & the old blackberries flashing & I keep out of the way. I guess she cant be overloaded with talent in the acting line. And lazy.

We still share the double bed, of course, & though there is still no question of the real thing, there is an unpleasant development which I hesitate to mention, yet I must.

In between what she calls her *flirtations*, when shes had a few drinks, she is as hungry for what she is missing as if it was food. And then she is inclined to insist for me to *see to her*, as she puts it (with hands, etc). I fight that, but then she shouts, If you dont I'll smash the flat up, so what can I do?

2835 James Street, Philadelphia
March 13, 1903

41st birthday.

Been in Philly since Nov. (on my own), Munyons sent for me to do 6 months managing for them here. No good pretending I'm not enjoying being the Grass Widower, & reading between lines of the few letters I dont think she's letting the Grass grow under her feet.

But I miss London.

34 Store St., London
March 13, 1904

And after a year, here I am – back in London for my 42nd birthday!

I was right about the Grass. Mr. Bruce Miller, a music-hall artist, Bass this time, with a Musical Saw. Havent met him, but she tells me he's American, tall & strapping, & she goes to him & his wife for duets. Its the one thing she seems to stick at, is the duets.

I asked what the wife was like, & she said, Hard to describe. You bet. Perhaps the wife turns the music pages for them.

I dont mind. She's sung at a couple of Smoking Concerts, guinea a time (Fancy Melba at a Smoking Concert).

Saw a letter on her dressing table the other day, *Dear Brown Eyes, Tuesday usual time, Bruce*. He calls them Brown Eyes, I call them Blackberries. Depends how you look at it.

Anything for a quiet life. And I notice when she goes for the duets, she puts on my 2 pieces of jewelry, plus the bangle.

March 13, 1905

I'm 43 today – and it's hard to believe she's still 11 yrs younger than me, every yr. she gains 2 & now she looks over 40. A rose in bloom you could say, though any day now the bloom will change to full-blown.

Bruce Miller leaving for the States next month for good, & she's been on edge ever since she heard, snapping snapping, so I felt a change would help.

So the big news is, that with the Store St. flat getting more poky by the week, Ive been looking around & (thanks to a Mortgage) we are *moving!* In 6 months, in Sept.

She agreed with me on plans & I spent a Sunday afternoon walking up & down side-roads off the Camden Road – a couple of miles north of Store Street & there I saw the house, with a TO LET sign. Just as I had dreamt it up – bay windows, three storys without the Area (basement part) & 2 trees in front garden, thick with leaves. Road dips down in graceful curve, with a slight slope.

Hence the name – *39 Hilldrop Crescent, Camden Town N.W.7.* Omnibus to work but worth it.

A big house, for 2 people. 2 sitting-rooms on 2nd floor (what in London is termed the 1st floor) with master bedroom & above that, bedroom & sitting-room, & above that, 2 bedrooms, also box-rooms (v. useful) I expected her to call the sitting-rooms *drawingrooms* – & she did!

In basement, kitchen & maids room, which we will use as dining-room, though she calls it the *breakfast-room*. (Ha!) On the ground floor, a little study for me.

Quite a mansion, but I favor a roomy home. And for an extra reason. For the first time since married, I'll have my own room, & *my own single bed.* (I'll insist on that, & I bet it suits her too.)

I called on the Landlord Mr. Lown, very pleasant, he said he would be relieved to have a Doctor for a tenant, 52 pounds 10 shillings a year per annum for 3 yrs. I am excited as a kid, she's not far behind. Last Sat I went for a wander up to the open-air Caledonian Market, very near Hilldrop Cr., where you buy everything from cabbages to parlor

suites, must shop there when we move in. Also, to celebrate move, buy her fox furs & muff.
Being handy with my hands, I'm going to wallpaper right through. Green I thought (restful) but she gives a shriek & says, With my Irish blood I'm superstitious about green, & green would be a Hoodoo on me in a house & bring bad luck.
I thought to myself – Irish? No kidding, I never knew Dublin was in Poland.
So it's to be pink, right through. *Pink!* The place will look like a whore-house, but if I insisted on the green & she so much as sprained her little finger, she'd blame me.
She had got Bruce Miller's photo by her bed. What a muddle of a woman – spends money like water on clothes – *my friends call me a Bird of Paradise* – but over Housekeeping watches the pennies like a refugee from where her Dad came from (Poland) says we cant afford a maid living in (all this room!) hates servants (& they hate her) she'll do the shopping & cooking & washing, which means she'll *half*-do it, hit or miss, & thats where I'll come in.
Anything for a quiet life.

39 Hilldrop Crescent, Camden Town, N.W.7.

Sept. 16, 1905

Not a birthday, but must record the imp. fact that yesterday I moved into my Crescent home – moved for life, I hope!
While that sounds as if I moved in on my own, the fly in my ointment is, of course, that I did not. And a big fly it is.
So last night was a housewarming, and too warm by half. First, last week in the Caled. Market I bought a very nice crystal chandelier & hung it up over the dining table (she having more than once harped on fact we must plan the odd dinner party for her theatrical chums, as she calls them).
Well, 6 pm yesterday the 2 of us are down in the kitchen, me peeling potatoes & her kind of laying the table with a gin on

the side, when theres a terrible crash in the next room. We run in and theres the chandelier in pieces all over the dining table.

Well she turns on me like a bat out of hell – you clumsy bastard, I might have been sitting under it, are you trying to kill me – nothing about what if her better half had been the casualty.

I walked out in the middle with the smashed chandelier in a dustsheet, and there it is on a corner of the cellar floor across from the kitchen, with sundry other rubbish from moving in.

A nice first night, says she, it means bad luck I just know it. And never opened her mouth again, except for food. And drink.

Not a good start.

March 13, 1906

44th birthday. Nothing else to report after 6 months here. Business not so good – & its one business after the other – its not that I cant settle to one job – I *am* reliable – but no job seems to be able to settle to me.

And not my fault, take the Drouet Institute, Regents Park, some months ago.

I worked hard for them, then I began to smell a rat – the man keeping the accounts had (to me) a shifty eye. Then both eyes got shifty, & it ended in the Police Court.

I had to give evidence, about seeing him practising signatures – oh how I hated the prospect of that! Sleepless nights, so I had to put something in my nightly Dr. Tibbles ViCocoa, to make me sleep at all.

To stand in the cold light of that court & swear on that Bible & then tell the Truth So Help Me God – she was in the Public Gallery & told me after, that even from up there she could see me shaking, I looked as if *I* had forged the checks.

Then the Sovereign Remedy Ltd. – the Sovereign turned out to be less than solid gold, went bankrupt. Then the Aural Clinic – failed last week.

Am in middle of reading *Diary of a Nobody*, life in the London suburbs. V. amusing.
A Nobody with a Diary, thats me. Better news next time.

March 13, 1908

My 46th birthday. After more than 2 years, I seem to be settled at last – my own set of rooms – a main office, plus my own private room – in a big block of Offices, Albion House, New Oxford St., where I work still for Munyons Remedies but also I work in partnership with 2 pleasant Doctors, Rylance & Masters (*Yale Tooth Specialists*).

I even have a *Dental Mechanic*, Bill Long who has worked with me on and off for several yrs, then theres young Taylor the Commercial Traveller who comes in & out, a bright willing lad that wants to be a Doctor, but unfortunately – like my Uncle Otto – cant stand the sight of blood, & in our profession, that is a serious handicapp.

We are all in harmony, & they seem to like me & my ways. I even cut their hair for them at times! (the Doctors & Bill Long) & sometimes (in fun) they give me the nickname of Sweeny Todd the Demon Barber of Oxford St. All friendly, though we are formal together, its always Dr. Crippen & Mr. Long etc.

And of course theres my Sec. & Book-keeper, thats been with me since the early Munyons days (8 yrs now). Since I work hard, thank God for her, I couldnt move without her, she is so good you dont know she's there. Quiet as a mouse & smooth as a machine.

A mouse-machine, that just about describes Ethel Le Neve.

4
LOVE

March 20, 6 pm

No not my birthday, only 1 week later, but I'm angry again & need to simmer down.
 Funny I should have mentioned my Sec. up above.
 With the 2 of us working quietly in my Office, Room 59, this afternoon the door bursts open. La Belle, in full swing, furs flying, great big fox muff (she had just come from an Audition at the Alhambra) glares at my Sec., slaps a paper on my desk as if it was a Writ, & storms out.
 Its a scrawled note, capital letters, DEAR MRS C. YOUR HUSBAND IS CARRYING ON WITH HIS SEC, ASK HIM HOW HE FIRST MET HER IN THE STREET, BE WARNED, A WELL-WISHER.
 Well, I look at my Sec., who is looking puzzled, & I tell her it's a Bill I had overlooked. (Dont want her to be worried. And I know who this is from, a Clerk in opposite office who wants to cultivate my Sec. & I have an idea she has snubbed him.)
 La Belle will be home any minute & unless Hell is to break loose, I must convince her theres nothing in it – ie. just tell her the truth!! Ie, how I met my Scarlet Secretary.
 Ie, that soon after we settled over here (in Store St., nearly 8 yrs ago) I was walking to the omnibus in the rain, on my way home from Munyons, when I passed a Pub & there were these 5 kids, several crying, waiting for their Parents to come out, the eldest girl seems to be the mothering one, looked 14 but was 16, I was so sorry for them I gave her a shilling to take them to coffee-stall opposite & warm up with hot drink.

Must explain to B.E. that I gave the poor thing my Office adress – because with it occuring to me that my prospects becoming good I would need a Sec., I decided to pay for the girl to take Shorthand & Typing – she was so painfully shy she said No, no, but she did, & of course took to it like Duck to Water, & grateful to me I'm sure but again too shy to express herself.

And never has – all she told me was that her Parents like going to Pubs (I guess a poor home life, her Dad working when he feels like it.) With the next sister (after her) getting old enough to look after the young ones, I advised her to move out into digs & she has (quite right) gradually got away from family.

And when B.E. is bound to ask why Ive never mentioned all this before, I'll say, Because dear you would have Suspected the Worst.

And again the Truth!!

March 27

Not angry this time, but in a flat spin alright.

That evening, was able to make B.E. see sense. But now I dont know where I am. I'm 46 yrs old, & just dont know. All of a heap.

This afternoon I was in the Office in Albion House – the usual business afternoon, dealing with wax-in-the-ears, filling come loose, etc. Rain, across New Oxford St. a regular pea-souper of a fog, so you could hardly tell an ear from a mouth, when – a knock at the door. A telegram.

I do get the odd business telegram & was just opening it when I saw it was for my Sec. I doubt if she'd ever had one in her life & opening it her hands were trembling. She looked like a scared child.

I knew why too – I'd overheard Taylor the Comm. Traveller chatting to her. The family live in a bad part of Camden Town – quite near Hilldrop Cr. – & the one she's really close to is her father. The telegram is to say he's been run over & dangerously ill in Hospital.

Well, she went dead white, looked across at me – she's the same height, & her face crumpled up like paper, & the next minute she'd flown over to me & her head on my shoulder & sobbing like a baby. There was a funny nice smell like of wild flowers somehow, it was the eau-de-cologne which on hot days Id seen her dab some on a little handkerchief.

Then she babbled something, twice – I couldnt get it, she was kind of hiccuping. And holding me tight.

It was a bomb-shell. I'd never so much as shaken hands with her, just Good morning Miss Le Neve, Good morning Doctor, & here I was with my arms round her.

This wasnt a mouse, nor a machine neither, it was a person. ('Poisson', no dont let me think of that.)

Well, I said, There, there, or words to that effect – nothing to stick in the memory – sat her back at her typewriter & poured a drop of brandy from the emergency cupboard. I said she must go back to her lodgings (Constantine Rd. Hampstead) & offered to see her to her omnibus but she insisted she was alright. I gave her money for her fare up to the Hospital & she went.

And left me in the Office. Alone. And I'd never been in it – or before, in the Drouet Office – ever, without her there. In her quiet corner.

A minute ago, as I was writing down the above, I heard that voice behind me – Hawley, what the hell are you writing, your life story? I said, I'm working on a treatise (as well I might be). But in future I'll confine writing to my private bedroom.

She's sitting there in one of her trailing negleeges or whatever, manicuring her nails. Seems to me that when she's not on show, her voice gets more and more of a rasp on it. And that swearing.

And to think I have got no idea what my Sec's voice sounds like, except for Good morning Doctor, yes Doctor no Doctor.

(Ive got to the end of the Diary, would you believe it. Because it is a Diary now, & Ive got kind of addicted to it.)

N.B. buy another tomorrow lunch-hour. A thicker one.

And I wont refer to B.E. by name, the name of Belle suits her less & less. From now on she'll just be she.

(Page 1, Vol. 2 of Diary)

Sat. March 28, 1908

Didnt sleep last night. And all morning, alone in the Office, not a stroke of work – I just sat looking at the corner where her typewriter was sitting, with a black hood over it like it was in mourning.

What does she look like, I thought to myself – and I couldnt remember. Slim, brown hair parted in the middle, tied in a neat bun, large eyes – but what colour? Neat ankles in black stockings – I had noticed that, & when she put her typed letters in front of me, small delicate hands.

What has she been wearing? Something dark, collar white. But I never noticed anything else. Ive been lying in bed in the dark, trying to picture her. Or staring at the typewriter – it was like trying to put together a ghost youd seen – but I'd been seeing this one for 7 yrs!

Then, still staring – all of a sudden I thought, Hold on & I went to her typewriter, felt underneath & pulled out a little drawer, feeling like a criminal or something.

Pencils, little bottle of our Liver Pills, old postcard from her Dad, a copy of the Sunday Companion, then the eau-de-cologne. I sniffed at it – & with the scent, that short minute comes back to me, her babbling as well. What was it she said, twice? And I remembered.

She said, *You* are my father now. Twice, she said it.

And I could have been too. When checking insurance or some such, I had once glanced at her Birth Certificate, & she was born in 1883, when I was all of 21. And now I'm 46, and her 25.

Thank God for pen & paper. If I hadnt got into the run – by now – of ocasionally writing things down, Id this minute burst.

Mon. March 30, 6 am

And Ive been like that since, 2 whole days.
    Couldnt sleep.
    When I got home from Office on Saturday, she was back from another audition.
    Blast them – *she* said – they wouldnt even let me finish my 1st chorus, & the bastards were only touring managers at that, that Mr. Didcot has the cheek to call himself an Agent, he wont be mine much longer.
    I looked at her where she was standing in front of the mirror, taking the hatpin out of her showy best (audition) hat with all that glitter on it (all that glitters dont mean contracts) glaring away at herself & fiddling with that dyed hair with the false curls, & I thought to myself, Shes living up to her name less & less every day. Belle my eye.
    Then I thought, Who is she?
    A stranger.
    And that slip of a ghost of a girl, who I only know from her typing, is close to me. She belongs to me, Im her father and her—
    Her what?
    I feel as if Ive been run over or something.

11 pm

Today endless in Office, me doing my own typing, like a crippled person stumbling around on a sheet of paper. I went & sniffed the eau-de-cologne, twice.
    Whats the matter with me?

Thurs. April 2

This afternoon, with me typing in Office, she walks in. A dark wisp of a thing, tired out & no colour.
    I thought, She's come from the Funeral and said – like a

dumb-bell – Youre back, how did it go, as if she had arrived straight from a wedding or some such.

She said, He nearly died but now off the danger list, thank you, then she comes right over to me.

I sat right back – what did I expect? – & she takes the heavy typewriter up, then back to where it belongs on her little table. She looks at my typing, tries to bite back a smile, takes up an eraser, puts it down again, rolls out my letter, and starts typing from it. She retyped the other letters as well. As I said Goodnight and she said, goodnight Doctor back, I thought I don't think Ive ever called her anything. Not even Miss Le Neve.

I know I'll be awake again tonight. This cant go on.

Fri. April 3

On way to Office this morning – walking down to New Oxford St. from omnibus – I bought a bunch of daffodils. Funny, all these years, with all the jewelry & new fur & such, I never once took home flowers.

I was putting them in the jug we have in the Office for hot-water for the tea – I like English tea – when she arrived, still in her black. Good morning Doctor – then she saw me put the jug next to her typewriter. She went quite red, which suited her better than being pale, & she said, Nobody ever brought me flowers before, thank you Doctor.

I said, I think by now you can stop calling me Doctor. Then I thought, I dont want her to call me Hawley.

And then I had a somewhat odd idea but didnt hesitate, just spoke out. I said, What's your fathers name? Peter, she said, I said, Would you like to call me Peter? She thought a minute, then she said, Yes I would.

Then we got to work as usual, me filling prescriptions etc & her fetching them to be typed. And I thought, I'd like to sit in this shabby room, just the 2 of us, right through tonight and all tomorrow, and so on. I dont want to go home.

What sort of a home could you call it?

Sat. April 4

I know I wont sleep, so here I am.

Tonight she gave a dinner-party, celebration. The great Belle Elmore entertains her select group of friends.

With not landing an engagement, she's been keeping in touch with the *Profession* by taking stage paper *The Era* every week, & via that, enrolling as a member of the Music-Hall Ladies Benevolent Guild (also through my old friend Dr. Burroughs being its Hon. Physician).

And so met this couple Mr. & Mrs. Nash (he manages her, she performing under name of Lil Hawthorne, a Variety Turn as they call it) then (through them) Mrs. Eugene Stratton, wife of a black-face comedian (American) who they say is famous.

Quite pleasant people, well-spoken for Theatricals – & the celebration was because she's been put on the Committee, fancy that.

Well, B.E. was straining herself to please, & that meant *talk*, non-stop, nobody could get a word in edgeways, the conceit oozing out of her like persperation. About her clothes – how she had to send the dress back twice before it was to her liking – how Mrs. Lucas the charwoman sneaks off with bits of butter & sugar, the women in the King's life, then laying down the law about some poor actress married to an Irishman, because the actress had asked him to divorce her, & what a sauce, she must know he's as staunch a Catholic as *I* am, & of course, he wont *hear* of such a thing.

I didnt say anything.

After dinner, she obliges with usual Song at Piano – the one she gave at last Stage Appearance (early last year, Bedford Music-Hall, 3 nights). *Down Lovers Lane*. (But not, since she was the great Hostess – *not* wearing the short spangled dress she had bestowed on her theater audience.)

It was never a big voice, & with the rest of her getting bigger the voice has got smaller. And quite a bit off the note.

Then whist. And the gin was out. She got fairly flushed & blotto over the cards, I always know because she doesnt bother (in front of guests) to hide her feelings for me. Ie, contempt. Youd never guess it was *my* jewels shes got on her!

Hawley, get an ashtray for Mr. Nash's cigarette & sharp about it, I'm in a draught, move the screen over, Hawley youve brougth out the sticky pack again, I *told* you—
The others dont seem to mind, but that voice goes right thru me.
I look at her & I think of that slip of a girl bent over her typewriter. Yesterday she called me *Peter – good morning Peter* she said.
She was wearing a costume with coat over hips, trimmed with braid and braid buttons to match, blue serge she said it was, matches her blue-gray eyes. The other one she has is a costume in what they call gray shadow-stripes. She dresses very plain & ladylike, just needs a little flash of jewelry to set it off. She's 100% perfect.

Sat. March 13, 1909

My 47th birthday. Nearly a year, and still 100%. One Saturday 6 months ago now (since we close the Office at 12.30 on that day) I asked her to lunch at the Holborn Restaurant. It's got to be a habit by now and every Saturday, my excuse to No. 39 is that I'm at the B. Museum Reading Room, brushing up my Medicine.
My Dearest (Ive taken to always calling her my Dearest, in my mind, & will henceforth refer to her as such, ie my D) is no easy talker, but Im not one to mind that – & when she *does* talk it's to the point. Sensible, wise & quite a sense of humor & she seems to appreciate mine, what there is of it.
She always loved her father of course, but with him being strict with her (& unreliable) she was always frightened of him – thats whats made her timid & not sure of herself. She says nobody ever did anything for her before I did. She already has confidence in me, & I feel that (with time) she will lean on me, completely.
Like me, shes never made friends, so this getting together (like two magnets) was meant to be. I count the hours till that Saturday lunch.
And Sundays are horrible.
I sometimes curse Bruce Miller for leaving the country, he

was able to keep the lady on a fairly even keel but she's getting worse & worse – I know her, she's missing it badly. Gin aint enough.

And two nights ago she got on to discussing finances – well it's true that business has been going down & down & Ive had to cut her allowance, & her idea now is that since its *bloody ridiculous* (British swearing, dont you know) for 2 people to be on their own in this rattling big Joint – with those empty rooms we should take in lodgers, we would find some as easy as pie thru the Guild, we'd get a good class of people. Plus an Ad. in the *Daily Telegraph*.

I dont mind, if it gives her something to do. And weve heard what taking in Lodgers can lead to, havent we?

Tues. March 23

That was last week – but whats more occupying my thoughts tonight, by a long shot, is what happened in the Office this afternoon. After 7 months of keeping distance!

The coldest day yet, this winter, & I saw her stop typing for a minute & rub her fingers, they were quite blue. I got up, doctor-like saying, We'll soon get the circulation back, & went & rubbed her hands together, quite hard, till they got good & warm.

Then I did something I still cant credit. I took her hands – palms up – & I kissed both of the palms.

Then I looked at her – there was such a beautiful look on her face that before I know it my glasses are in my hand & weve got our arms round each other, & kissing. Properly.

Well we stared as if we'd never seen each other before – like in a panic. Except *I* couldnt really see her, without my glasses she was just a fuzz.

She said, First time I've seen your eyes, theyre nice, theyre gray. I said, I've seen yours, theyre nice too, & bluey gray. She smiled & said, Why dont you put your glasses back on & take another look?

Then I said, You dont mind the glasses? She said, No,

theyre a part of you. Well (I said) I have a few false teeth as well. She said, quite natural, with a smile again – I know, Ive seen you cleaning them, thru that door (Transparent).

I said, I would have thought youd have liked young Taylor much better than me (the young Commercial Traveller). Then she says, He doesnt need glasses & I'm sure he's got all his own teeth, but I dont like him the same.

Then we stare at each other again. Luckily it was near time to go home, so I snatched at my bowler hat & coat & said something like, This wants some thinking over, & went.

And I never spoke a truer word. Ive been thinking it over ever since.

Wed. March 24

I realize now, that all my life Ive been kind of drifting – letting things happen, easy come easy go, anything to avoid trouble. And what *makes* me realize that, is that although I'm sitting here in my slippers, & pretty well shivering with the early-morning cold – in spite of such a handicapp, for the first time in my life, I am sure of something.

And that something is – that I have got to have her – not just in the crude manner, I wouldnt speak of her in such a way – but to have her with me, for the rest of my life.

Yes – at long last – & about time too – I understand what all the poetry & ballads & stuff have been going on about these hundreds of years – Drink to me only with thine eyes, I want you for my very own, till the Sands of the Desert grow cold, all that stuff.

In my humble view now, all that doesnt go far enough.

Thurs. March 25

In the Office today, nothing more said – just a warm feeling in the cold air, Whats the word – nearness, thats it. We understand each other.

An hour ago I was sitting staring into the fire, with her (the Other One) opposite in one of those negleege things – stained & frumpy by now & the old glass of gin in her hand, when she says, all of a sudden – Youve changed.

I look at her, quite startled – have I? Youre all absent-minded she said, youre thinking of something else all the time. I said perhaps it was Anno Domini or some such, & she said, Oh, no, if anything youre more sprightly than you used to be, nipping up & down those stairs.

I said to myself, My friend youre on thin ice. But felt quite complimented just the same.

Fri. April 2

Nothing more said in Office, but my minds working.

After leaving Office this afternoon, I called at the big Walpole Hotel in Tottenham Court Rd., to see a Mr Jesson who runs it.

Last year (as a sort of extra-mural medical activity that I got into through another patient) I was able to help him over a situation where discretion was needed – *viz*: venereal disease, & I know he's been grateful to me since.

Last evening I asked him for a little discretion in return – *viz*: the use of a room in his hotel, every Sat afternoon, 2 till 4.

Its booked for tomorrow week, Sat. 10th. Provisionally. I wont bring it up till lunch that day at the Holborn. What if she gets up & leaves the Restaurant?

Sun. April 4

The lodgers are in. 3 German students, nice fellows, plus Richards, a sturdy red-haired Cornish fellow (more lower-class) who is nightwatchman or fireman or suchlike at the Avenue Theater.

And therefore in the house of a daytime.

Yesterday evening the fireman says to me, quite jovial, Doctor what about getting the Christian names going, Im Jack what are you? Without thinking – Peter, I said, like a shot. And Peter I'll be. I looked forward to her hearing them call me Peter & getting mad, which she did too – Whats this Peter business? – I told her it's a third name I have & I'm sick of *Hawley*. But every time she hears *Peter*, she growls & flashes the old blackberries.

Sun. May 2

It's time I recorded that I said what I had to say, & my D didnt leave the Restaurant.

When she did leave, she left with me, & we took a cab to the Walpole Hotel. And have every Sat. aft. since. Yesterday was our 4th.

That hotel room, with no picture nor ornament, is a home to me such as 39 H. Cresc. has never been. Because these afts. are, from the start, so easy & peaceful – I never dreamt I would gain such happiness & I dont know what Ive done to deserve it.

I suppose its because (unlike some) she doesnt make demands nor sarcasms that would frighten even Casanova off – she just loves me being close to her, & of course that leads beautifully to the rest of it. And no pajama-top neither!

I'm a new person. At 47!!

5
HATE

LIFE AT No. 39 is strenuous. With Mrs. Lucas only coming twice a week, theres a lot of work & I have to get down to it before the Office – cleaning boots, making breakfasts, then the beds. But I'm up early anyway with my thoughts, & the chores keep me from brooding.

And Sunday is less of a nightmare – I play cards with the lodgers, makes life easier.

Mon. May 31

Miss Belle Elmore, Hon. Treasurer of Music-Hall Ladies Benev. Guild, has persuaded the Guild to move their premises to Albion House, just above my Office & Drs Rylance & Masters, Committee Meetings every Wed. She's got a nerve.

Is it to keep an eye on me?

Tues. July 27

No, just to tease me – during 2 months she has never come near my door before or after a Guild meeting. Thank God, couldnt bear to see her face to face with my D.

Wed. Sept. 1st

Thank God for the Saturdays.
Same feeling, only stronger. The only part I cant bear is

when I look at my watch on the bedside hotel-table & it says ¼ to 4 & I see the look in her blue-gray eyes.

And I dread seeing her on to her omnibus to Hampstead, to her digs – but its a comfort to know that her landlady Mrs Jackson is fond of her & treats her like a daughter. I dont blame her. We love each other, for ever.

At No. 39, *she* snaps at me now & then, but not as bad as before. Jack the Fireman seems to have done the trick. Hes welcome to all the grunting & perspiring.

Tonight, she had the Nashes to dinner. (She the Variety Turn, he her Manager). My Lady Belle alright till she gives a big sigh & says what she needs is a change. I nearly said, What about our Jack isnt he a change? But she meant a holiday & I knew why. The Nashes had just come back from a week in Paris. Oh I would like to see the Continent, she said.

I reminded her that Mr. Nash had just been left money by an uncle & I hadnt – but as to the Continent, I might scrape together enough for a day-trip to Dieppe.

Quite neat, I thought.

I notice in the paper they are advertising this stuff *Harlene for Hair*, supposed to cure baldness. Must get some, try anything once.

Thurs. Sept. 9

What I knew would happen, has – she's got sick of the lodgers.

J could manage the early mornings – breakfasts, boots, beds etc – but there was a lot of work somebody had got to knuckle to during the day, & she turned against that, within a week. Most evenings, by the time I got home, she was fit to be tied.

Also I got the feeling that Jack the Fireman (during the day) is more inclined to snooze than he was on first moving in, & this circumstance has done nothing to improve her temper.

Anyway, all four got their notice & went yesterday, leaving the Crippens on their owsome. 2nd Honeymoon.

Wed. Sept. 15

Some Honeymoon!
Back to the rows & hystirics. She vomited last night. I go to bed worn out, disgusted.

Thurs. Oct. 28

Life goes on, after 6 weeks nothing worth writing down, no plans to work out on paper. I live for Sats, but I worry about my D fretting about the future – its no life for a young person of 26, & I know her mother has been writing her about getting married – I dont know if Ma knows about me, she may, theres always somebody to gossip.

And life at home gets worse & worse – no good saying *at home*, its no home, just No. 39 – & her more of a slut except when theres company.

Mad on dressing up, & still got the idea one day she'll be a big star, so her bedroom – whenever I go in there to fetch something for her – is like a second-hand dress shop, ostrich feathers lying about, false curls, sequins. And everything smelling of smoke & gin.

Ive tried to get her to apply for Vaudeville engagements, but she never clicks and gets more bad-tempered than ever. Thank God for the Nashes, & now the Smythsons & the Martinettis, those last a retired Act (him a juggler, fancy a Retired Juggler) – she had met them through the Guild of course, a quite sprightly couple that seem to be amused by her, & that makes her take to them.

I just sit there while they talk *Shop*, oh the fun we had once the Rag went down (meaning the Curtain), arguing about which Comic has the best gags, getting the Bird, Front of Cloth – it's double dutch to me. I just sit there.

Not that I dont like the Music-Halls – last week I invented business dinner & took my D. to the Palace Th. and there was Julian Eltinge, a clever American chap, quite hefty, dressed as a woman with evening dress and ospreys. Now my D. and I never mention la Belle Elmore by name, but I could not resist whispering to her, Thats the woman I married. It was so sudden & comic that she laughed right out loud.

Fri. Oct. 29

Amusing occurrence Albion House this afternoon. Business slack, so was persuaded by that silly Miss Curnow to spend 2 shillings along the passage with Mme. Celestine, Clairvoyant & Horoscopes, retired singer in Operettas, just moved in. Long robes and bandeau across false hair, French accent ditto. I told her my birth date – March 13 – then a lot of abracadabra and study of my palms, long lifeline, etc., a lot of rubbish I think, anyway who wants to be in a wheel-chair aged 90?

Then the Horoscope – oh she says – Doctor you were wizz your Sun in Pisces, which means Water, but your Moon was in Leo, which is the direct opposite, Fire (I never heard such nonsense) so it means, Doctor, zat you are *very* contradictory character – (quite roguish she gets about this) – the Pisces half of you is modest & retiring, you are one zat enjoy being at zee back of zee crowd.

Thats quite a coincidence, when I remember what I wrote 7 months back, that I'm one to avoid trouble. Then she says, now zee Leo half is zee opposite, it is zee half zat stand up & say, I am in zee right & nobody else is, nozzing will change my mind, I am center of zee stage. (I thought of having seen Caruso in the box & the worship for him, & again noted the coincidence.)

Then I said – careful not to cause offence with scoffing – And what, Madam, does that add up to? Oh, she says glib as glib, it can mean zat zee person ees leading a double life!

She looks at me all coy again, & for a second I think, Has somebody been gossiping? No, just another coincidence. All rubbish really. Waste of time & waste of 2 bob.

Tues. Nov. 2, 6 pm.

I knew it would happen, sooner or later. After 7 months, it has.

Because of which, I have now got to face a 2nd talk with B.E. when she gets home from a Fitting or whatever it is. Just got home to find she's left a paper on my desk.

DEAR MRS C., *NOW* WILL YOU BELIEVE HE IS CARRYING ON WITH HIS SEC, SAME WELL-WISHER.

I'll now get my answers ready by writing down what I should say – it'll help me—

Now be sensible dear, every wife gets that sort of note sooner or later, theres always enough mischief-makers to go round, as a matter of fact Miss Le Neve is young enough to be my daughter, & anyway is engaged to be married, though being a nice reserved girl she doesnt tell everybody, every man with a secretary that he appreciates is accused of this sort of thing, it's a joke by now, anyway I'm not the philandering sort, *you* should know.

(Thats a good one, remember to use it.)

Here she comes—

Wed. Dec. 15

Tonight usual drunken hystirical abuse, & on top of that something new, & she better watch out or I might do something I would regret.

The evening of the anon. letter, 6 wks ago, she didnt know whether to believe me or the anon. letter, but she seemed to decide to ignore it. Well, tonight she brings it up, & starts on my Dearest – dont you try (she shouts) to pull the wool over my eyes, youre messing about with that little whore & one of these days I'll walk down from the Ladies Guild to your Office & face her with it, the brazen bloody trollop.

That from *her*!!

I hate the ground she walks on.

I hate her. I hate her.

Tomorrow, on my way to Office, will pick up Princess Robe, with sequins, from Bradley's. My D's Xmas present.

1910

Sat. Jan. 1

New Year's Day & what a New Year's Party *we* had! She got drunk, behaved herself till theyd all gone, then went berserk, smashed a chair & 3 glasses.

I cant stand it – Ive *got* to give her something in her drink to calm her down. Must call at Lewis & Burroughs Chemists.

Burroughs? Same name as my old friend Dr. John Burroughs! The coincidence may bring me luck.

Fri. Jan. 14

Another row tonight, & of course my D's name dragged in the muck, top of her voice. Then vomited.

I hate her—
I hate her
I hate her
I hate her
I hate her
I hate her

I feel better after that. Dear D – & for a change I mean Dear Diary.

Wed. Jan. 19, 7 am.

Beginning of last evening she was quite perky, had heard a rumour at the Guild that Bruce Miller (her American Bass) is coming back to try his luck once more in London. (I sure hope he makes it.) Later, with the gin, she got to making bedroom hints & I only just managed to side-step that which makes me sick to my stomach.

Calling today Lewis & Burroughs to sign for Hyoscin & pick it up.

Hadnt heard of Hyoscin in England, but thank God I remembered that its used in the States, by the people who know, *in cautious amounts* to calm patients down who have problems of temper due to Alcohol & sudden sexual urges. (NB Bear in mind that correct procedure is to reduce crystals to a liquid, then to be used in v. small tabloids. Prescribed dose, one 150th of a grain, watch that.)

Mr. & Mrs. Eugene Stratton to dinner tonight, with him being the well-known comedian she'll get excited & the gin will be out.

MEMO for plan, for each time she needs calming down. Slip up to bathroom, slip bottle into my jacket-pocket, back downstairs again carefully slip exact dose into gin.

Thurs. Jan. 20, 7 am

And last night it worked!

I saw all the bad signs on her when, as the Whist comes out Stratton remarks, really jovial in his stage manner, Doctor what about getting the Christian names going, Im Gene what are you?

Peter, I say & as she deals the cards like shots from a gun, she gives me that look, another shot from a gun.

So I made the old excuse to slip upstairs & when I'm back and getting her next drink, Dr. Crippen administers the exact medicinal dose & in 15 mins its acting like a charm.

In the middle of biting my head off she trails off, and gets all drowsy. Not enough to draw attention from Strattons (theyd seen how partial she is to whistle-wetting) – she seemed just mildly squiffy & kind of lazy.

And once theyve gone, she just yawns and goes up to bed like a lamb. I take that back – goes up to bed like a big stout

staggering sheep. Thanks, Lewis & Burroughs.
Longing for Sat.

Sun. Jan. 23

Yesterday, at our Holborn Rest. lunch, an important moment.

I was making light of having to get my own breakfast & clean my own boots, even after the lodgers have gone – when she says, But Peter dear, doesnt – isnt your breakfast prepared for you? (As I said, We never mention her name).

Oh no I said, the lady of the house is *not* an early riser.

And then I said, Except on Sundays, of course, when she goes to Mass in Kentish Town.

I could have kicked myself. Too late. She said, She's a – Catholic?

Yes, I said, I go to church with her sometimes, to keep her quiet.

She talked of something else, quick, but I know what she was thinking. Marry a Catholic, & youre jailed for life. For the first time, there was something a bit sad in the Sat. afternoon.

And it set me thinking.

10 pm

This is an important entry, the most so far. In the last few hours, I have been thinking hard. And it gives me a big kind of relief to write it down.

By this evening she had been drinking all day, not heavily but enough to nag, & several things to turn her sour – a letter from the Teddington Music Hall cancelling her 3 trial nights in March, Mrs. Lucas breaking a cut-glass vase that had been a present years back from Bruce Miller, etc.

Well, she was sitting there sipping & trying to read a fashionable novel when she asks me – or orders me, more like it – to refill her glass. When I relieve her of it I realize she hasnt had a bath since Wednesday morning, ie attention to personal hygiene is no longer an important factor. Like most doctors I am squeamish on that, wont go into details even on paper to myself.

Then, just to say something, I suggest asking the Burroughs to dinner (the old friends from Store St.) I havent seen him for a long time. Oh no, she snaps, Ive turned against the both of them & I wont have them in my house. Then she sits picking her teeth & skipping through her Marie Corelli.

Then I thought, how dare she call it her house, its not her house & I want her out of it, I hate her I hate her—

And then, with her turning a page with a sniff, I felt something click in my head – like a watch breaking the silence & starting to tick, & isnt going to *stop* ticking, neether. And I said to myself, so clear in my mind I thought for a minute Id said it out loud – I said she's got to go.

And inside myself – slower – I said it again. In those couple of seconds while she turned that page, little could she guess what big seconds they were.

She's got to go.

6

I'VE ALWAYS BEEN TIDY

Mon. Jan. 24

SINCE LAST NIGHT, I've been thinking again. This has always been a Secret Diary, of course – what you might call innocent secrets, all the way. But from now on the secrets are secret-secret. Dynamite. (I ought to give it up, of course, but too late now, I need it.)

In front of me Ive got a small strong-box, bought it yesterday to keep the thing under lock & key. (I'll destroy it when the time comes, dont worry.)

Ive always been methodical in my work, all the same items catered for, & I'm going to be methodical now. Tidy. Plan of campaign.

Means telling fibs to my Darling, I dont like that but it's got to be done.

Let me think. A little sea-trip necessary.

Dieppe. Didnt I make a joke about maybe a little holiday there, one of these days?

Tomorrow Ive got to get this thing started, & got to word it right. And as Ive done before, I'll use this Diary for a quick practice of suggestions & explanations I may have to make next day. A sort of Memo of Future Talk.

MEMO of Talk, tomorrow at Office. Good morning My D, got a letter this morning to No. 39 – from a Monsiur (cant spell it) Larive in Dieppe, big patent-medicine firm, suggesting meeting with view mutual agreement, wants me to pop over on night boat Newhaven-Dieppe this Thurs, will be back (of course) for our secret Sat.

Report same to No. 39, only careful to say letter arrived at *Office* & I left it behind *there*.

Fri. Jan. 28

Back from Dieppe, where I half-expected my Frenchman to meet me in! Had quiet morning in a little cafe & practised my French, it seemed to come back, I wished my D could have heard me.

I even had a glass of red wine, since my little job was done.

And the little job had been to stroll carefully round the boat at about bedtime & take stock. Railings everywhere, of course, but right at the back – the Stern I think they call it – right next to the railing a little steep stairway leading up.

Now if you were told that if you stood half way up the stairway you could already see the lights of Dieppe miles away – lovely sight – you might be tempted to step up to have a look. Quite slippery, those steep stairs, & there's quite a noise from some machinery near by, goes on all evening, thats why its so deserted.

It'll be Feb. of course, & could be a rough night, but with any luck it wont be, Dover-Calais is the rough crossing & I believe in luck.

I enjoyed the wine.

She's not home yet from the Guild, so I can now get my speech ready.

MEMO Talk, at 39. Yes dear, excellent business deal, & what do you think – Monsiur Larive has asked us both for next week-end, so youll get your little holiday after all, the only thing is theres been money & jewelry missing on the boat, the Steward said its common on those night crossings, so they strongly advise ladies *not* to travel their jewelry on those trips.

On 2nd thoughts Monsure suggested a *midweek* visit – the Wed. to Fri. (I put that into her head so as not to cut into our Secret Sat.)

She's coming up from the garden gate. I can tell from her walk she's in a bad mood.

10 pm

I was right – she got back pretty grumpy as usual, & grumpier still when she found that while she was out Mrs. Lucas had cracked a mirror – bad luck, bad luck – but when I explained about Dieppe she got quite interested, & all arranged.

Will make a change of air from all this, she said, she's quite flattered but wont admit it. Was pretty miffed about not taking the jewelry but saw the sense of it.

Martinettis coming dinner and whist Monday, she'll be telling them all about her invitation to the Continent.

Everything going according to plan.

Mon. Jan. 31
or rather Tues. Feb. 1, 1.45 am

Worn out.

All fine through dinner, she had just enough gin to 'sparkle' – 'they said I sparkled last night' – then dinner over we get the usual rendition at the piano – this time my least favorite of the Repertoire, as tried out some years back at Camberwell Music Hall – *She Never Went Farther Than That*.

Then all of a sudden Paul Martinetti (in all innocence) mentions that the Smythsons have heard from her Bass (Bruce Miller) that he is *not* coming back after all.

Her face goes from red to black & back again, they couldnt help notice, & she didnt speak for 5 minutes. (I was thinking, Cant he send a pal?)

Then when we got up, Paul M. says, Well Peter you must be pleased about this Dieppe plan – & I knew that him calling me *Peter* would make her madder than ever.

Then he aks to be excused (to go upstairs & wash his hands) & five minutes later we hear a crash & he comes limping down. It had slipped my memory (a) that the gas mantle on the landing was out of action & (b) that the lav. window was open (on a winters night, & Paul M. already had a cold). He had missed his footing in the half-dark & was wheezing (a bit of a self-coddler if you ask me) & only a twist of the foot, not even a sprain – & he *did* say, its nothing Peter. Nothing.

Well, she turned on me & screamed. Wheres your bloody manners, letting a guest go upstairs by himself, could have broke his leg for all you care, and caught pneumonia – there are times when you disgust me Mr. P.T. & you know it, etc. etc.

Lucky the Martinettis were such friends of hers, but even they didnt know where to put themselves, & Paul (trying to make light of things) said, What does P.T. stand for, & then I (to stop her telling them it meant Pajama-Top) said its a teasing way of saying *Pete*.

Which (I must say) was quick of me, considering that (inside) with her bringing *that* up, I was seeing red. (Funny, they spell it *pyjama* over here – well however they spell it, I was real mad.) Then the gin started flowing & I knew I was in for a bad bad night. And that it would end with, Come on, youre a washout but a hubby is a hubby, do *something*—

I ask the Martinettis to excuse me – Paul M. calls out, in fun, Dont *you* fall over – I go up to half-dark bathroom (gas on low) grope at back of cupboard shelf for bottle, get it into jacket-pocket as usual go back down, take her glass for refill & slip into it required dose. Thank you Lewis & Burroughs.

At 12.45 I went looking for cab for Martinettis, luckily it wasnt raining but even so it took me ¾ of an hour to get one & when I got back she blamed me for the delay, but she was cheerful with them & called to Paul M. to wrap up against the cold which Id helped to make worse. They looked quite sorry for me. (All her friends like me, I dont know why.)

Well, once theyd gone she really let me have it, picked up an ashtray & it whistled past me, could have gashed my face. Then she rampages upstairs stumbling & muttering which (I was hoping) meant that my medicine was beginning to work.

Then I hear the bang of the bathroom door, & knowing how bad the light was in there, I knew she was falling about – well, I could *hear*. Then she lurches out & yells to me,

What youve done is to give me a splitting bloody headache, & into her bedroom.

I go up to bathroom, & my guess is right – she's vomited in there. I turn up the gas, hold my nose with one hand & set to with the other to clear up after Belle Elmore the Sweetheart of the Show Business.

Then, at last, up to my bed. She's at her door again, screaming after me – My headache's worse, get me my headache pills from the bathroom.

But I was fed up, & went to my door & called out, Get your pills yourself. Then sat down in my room to write this. I could just hear her stumbling to the bathroom for same headache pills.

I hate her. I h.h. I h.h. I h.h.

Before I started writing this, to remind me of other things, I thought of our trip to Dieppe next week. Its not long to wait. My D., everything going according to plan.

6 am

No its not going according to plan. When I wrote I was in for a bad night only 4 hours ago, it seems like days, little did I—

What do I do *now*? Tell me what to do – yes, *you*, Diary.

To think out how I stand, I must write it all down, carefully, it will steady me—

An hour ago I was woken up. She'd called out shrill, twice – I thought, Oh God she's going to vomit again – put on my robe, ran into her, lit gas & looked down. She hadnt bothered to get into nightdress & was sprawled naked on the bed.

But she wasnt wanting to vomit, there was something else in her throat. A dry rattle.

I rush back to my room, light gas, take my jacket off chair & take out bottle of Hyoscin tabloids. It *isnt* the Hyoscin, its her headache pills – & I remembered that with the gas low in bathroom, & the mantle outside broken, it was hard to see – I had felt for the bottle in back of medicine-cupboard, & *taken the wrong one.*

I'd put headache pills in her drink, & *she* had staggered into the bathroom (in the half-dark, sozzled) & taken the hyoscin. Probably 100 times the safe dose.

I run back to her. You just know when somebodys dead. She was dead.

Then my thoughts start going flash-flash. First flash, is back to Salt Lake City, seeing Charlotte on that bed – it's an object, like furniture, like a big bare doll left lying about waiting for a kid to dress it.

Then another flash – she's gone, and me and my D are all right, for ever & ever—

Then another flash – but its here, in my house – I wasnt even calling it *she*, I was thinking, *it*.

And I'm responsible.

I'm confused, & yet not. Must get it right – it was an unfortunate accident, of course it was – but if it *hadnt* happened, this body – by next Thursday, with luck – this same body would be rolling about in the English Channel & I would have got it there. So I cant feel anything different from feeling this is my doing just as much as next Wed would have been. Its like planning to catch a train next week and finding yourself already on it.

The difference being – Ive got a problem here that I wouldnt have had next Wed. It's *not* rolling about in the sea, it's in that bedroom with inside it a massive overdose of a poison which I bought in person from a known chemist, & signed for, they could check on that within the hour.

How could I ever *prove* it was an accident? With luck her fingerprints may be on the bottle as well as mine, but how could I ever convince them that I didnt advise her to try it, knowing that was good for her, being a Doctor?

My life would be over. Then they find out about my Saturday afternoons – they always do – no, I couldnt bear my Dearest to be dragged thru all that.

No, here I am in the middle of the night with the whole of London asleep all around, & only me awake. In a desparate situation.

What do I do with it?

Desperate remedies.

First, work out tomorrow – no my God its 7 am. it's *today* – Gen. plan. Must (at all costs) turn up at Office this morning at 9.30, and therfore keep regular hours within reason – everything as normal as poss, for future reference, if ever necess.

The work at No. 39 to be considered as a nasty spring-clean job to be done as quick & neat as poss. Like when a roof's fallen in & you have to clear away the rubble & then sweep up.

Work at No. 39, 6 pm to 11 pm, and then knock off (Light in kitchen wld be suspicious any later). Up again soon as it's light. Wear only under-drawers. Move hip-bath from bedroom down to kitchen.

Details for rest of today. Lunch-hour, walk down to 1 King Rd. Mansions Shaftes. Ave, to ask how Paul Martinetti's cold is. And tell them that when I left No. 39, she was still in bed. Never a truer word.

Throw Hyoscin away. Wash & dry her glass. Check there are enough candles for lantern for cellar.

Get out of bottom of trunk Cunningham's Manual of Practical Anatomy I bought in Charing Cross Rd (all those yrs ago!) Take it in brief-case to Office for close study during day.

Second thoughts – mustnt dawdle at Martinettis, just in & out, spend rest of lunch-hour shopping. That will make me late at Office, will tell my D that Martinettis kept me.

Shopping list. (Already have spade for clay from garden – NB, Not more than one Item from each shop.) Small axe, saw, pickaxe, big waterproof sheet, quicklime (2 sacks).

(NB, Before each shop, take off glasses. Dump parcels of shopping in Albion H. basement lumber-room, to be picked up on way home)

In No. 39, sharpen carving-knife.

Food. Theres fruit & vegetables in larder & I'll shop for more. (I most likely wont be hungry, but if I do get peckish, I wont be fancying anything cooked – I'm not squeamish but its only natural)

Must keep going. Gruelling, but I'm stronger than I look. Whats got to be done, got to be done. And any time I get tired & fed up with it, I'll think of my D, she must come to no harm.

The future is ours, & thats all that matters.

I'll put my head in at the bedroom door so to make sure everythings OK, put a sheet over it, make myself a cup of tea.

Give hands & fingernails final scrub before putting on stiff collar & frock-coat for Office.

Brain clear. Off to Office.

6 pm

Just peeked in bedroom door, its still as I left it (where else, but you have to check). Under the sheet, like in the mortuary.

Then I had a funny impulse. I telephoned my D (she was still at Office, working late) & told her I loved her. Oh she said, quite worried, Are you sure she's not overhearing you?

Quite sure, I said. Which was true!

All went to plan during day, was only ½ hr late from lunch. Just going to unpack shopping.

But first, before its dark & while Im fairly fresh, remove *it* down to kitchen. All those stairs!

½ hour later. It was good idea doing that while fresh – the *weight*! Like a sack of potatoes, had to roll it down the stairs & along passage. Never realized stairs & passage so dirty.

One v. bad moment – rolling it down the stairs I slipped, my glasses fell off & rolled under it, I thought I had broken them & that would have been terrible. But God was on my side.

MEMO – must bring spare pair tomorrow from Office.

Was quite out of breath, made myself a cup of tea.

Now refresher notes after study of Manual of Anatomy. Funny how these things come back to you, after yrs. Then memorize notes.

Notes. (1) Commence dissection of Thorax *after* upper limbs removed from trunk. (2) Hip-bath from upstairs for drainage of blood & semi-liquid bowel content. (NB. For

this must have nose-protection – I just remembered clothespegs in kitchen drawer, sounds silly but when I clipped one to my nose I found it worked!) (3) Backbone needs fine-toothed wood-saw, easier than long bones (arms, legs) so tackle latter first. (4) Impractical to try burn viscera since viscera wet – ie heart liver bowel pancreas spleen – so set aside for burying. (5) Memo – no good tampering with skull as liquid wld flow out, ie brain-matter soft like junket so keep Head intact & dispose later.
Now the hard part.

Wed. Feb. 2, 8 am

Worked till 11 pm, was so dead I slept till 6. Woke up with light, worked. Its laborers work at that, yet here I am, fresh as paint, & preparing my Memos for the day! Something is keeping me on the go, like some sort of clock inside me is wound up & driving me.
 A minute ago, a bit of a scare. A knock – milkman. And me in my under-drawers, & not presentable besides (ie not clean).
 I nip into cellar, he knocks a couple of times & leaves the bottle. I was glad of a glass of milk, though of course I am making tea quite a bit, also Bengers Food, v. nourishing. (Cellar floor made up of irregular broken bits of street paving, plus bricks, a couple of those loose – a bit of luck, if it had been a smooth floor it would have been imposs. to cover everything up after the job without it being clear theres been disturbance. As it is, by the time I've put back the bits of coal & swept back the coal dust and put that unlucky old smashed chandelier on top of the rest, everything looks exactly as it was.
 All I remember of my own feelings during the nights work – as far as I could have any feelings, the work was so tough & detailed & you had to have your wits about you every second – was feeling so alone at moments (as alone in the middle of the night as in a disused coalmine) that I kept longing for company, such as one of those H.M.V. phonographs (called

gramophones here in London, just to be contrary) with me winding it up & Wilkie Bard in a comic song, or Vesta Victoria, *Riding on a Motor-Car*. And what a comfort it would be to have the dog around that listens to His Masters Voice, he would be company too, except awkward with him scampering up & down the stairs sniffing & wondering what the deuce was going on. What outlandish ideas run through the head when under stress.

MEMO of my talk which must follow this morning with my D in Office.

My Dearest Im afraid I have got some news, get ready for a shock. When I got home yesterday eve (Tues) I found she had packed a bag & left me, for America.

You see, the night before, we had had the final blazing row & she said she was leaving for good, to join Bruce Miller – you know, her old flame the American Vaudeville chap. I wasnt sure if she meant it so didnt mention it to you yesterday, but shes gone alright.

She must have packed in tearing hurry, didnt take half of her things, left all that stage finery & rubbish. Train to Liverpool I suppose, I dont even know what boat.

Did you ever know of anything so mad, Im upset with the humiliation of course but the house is so quiet you wouldnt believe—

Today being Wed, she'll be missed at the Music-Hall Ladies G. Committee Meeting, so 2 notes for my D to take up to them. One to Sec Miss Melinda May, 2nd to Committee, in my writing, signed Yours etc Belle Elmore, per H.H.C.

Dear Friends, Please forgive a hasty letter, but—

(No, I cant give *them* the idea she's walked out on me, make me look a fool, how can I go out of my way to be the laughing-stock of all those Guild gossips? I'm made of better stuff than that—)

But I have just had news of the illness of a near relative & at only a few hours notice I am obliged to go to America. Under the circs I cannot return for several months, & therefore beg you to accept this as a formal letter resigning my Hon. Treasureship. I am enclosing cheque-book &

deposit-book for use of my successor, etc. I hope some months later to be with you again, & in the meantime wish my good friends & pals to accept my sincere and loving wishes for their own personal welfare.

Believe me, Yours faithfully, B.E., pp H.H.C.

I'll tell my D that the reason the letters are in my writing is *she* was in such a tearing hurry she asked me to write them for her.

For a chap whos never done it before, Im getting to be quite a good liar.

11 pm

Work O.K., worst over. Tomorrow eve will be mostly clearing up before Mrs. Lucas's day next day (Fri.).

I'm tired already – its the digging that takes it out of you – yet cant stop writing.

And perhaps helped by what I have in front of me. A glass of her gin!!

Tomorrow morning, pack her big empty suitcase with stained clobber (towels sheets newspapers etc to do with clearing up) plus all the things she *has taken with her to U.S.A.* – all her make-up muck from her dressing-table, brushes etc, several pair of what they call *lingerie*, a couple of costumes, shoes, stockings etc, passport, handbags, couple of hats. Stuff into same suitcase the pick-axe, & – most important – something which in the meantime I have in cupboard wrapped in towel – the head. (I remember that a surgeon told me that 1st London trip, that a head seems to weigh as much as the rest put together. And by golly its true.)

Wrap it in all the costumes & lingerie, pack the lot into suitcase & place suitcase in back of box-room to be disposed of in good time. (Luckily, attic temperature v. cold.)

Also that painting of her – thats for the attic, those eyes follow me about & put me off.

All went well at Office today – I seemed to get a 2nd wind – except my D gave me quite a turn. After she read the 2 letters which she was to take to the Guild, she looks up at me, in the eye, quite quiet, & says, Peter why cant you tell the truth?

I looked at her, thinking for a minute she meant – but of course she meant about my wife walking out on me. I couldnt make her see that I couldnt tell anybody (except her) that I had been humiliated, but she couldnt see it. Women are funny.

Fri. Feb. 4, 7.30 am

All done. Present & correct. Have packed furs and jewelry into suitcase to take to Albion H. for my D. Must get there early before she does.

7
OVER AND DONE WITH

LAST EVE (THURS), swilled out everything, then went over house as if with microscope. And basement clean as new pin for Mrs. Lucas today. But I wouldnt be a butcher for anything.

Strange that I was more disgusted, that last night, by her vomiting than by anything in the last 3 days.

Over & done with.

Being a thoughtful man, I am considering how I am feeling at this moment, & am releived to note that the cellar is not going to be that much on my mind. To put it bluntly, to me it is no different than her having been put to rest yesterday, Thurs. Feb. 3 1910 at 6.30 pm (exact time!) in Kensal Green Cemetry, Dearly Beloved Wife Of. Only less expense, no tombstone etc!

And imagine a cemetry – we've all seen cemetries, from the top of an omnibus or whatever. And theres always houses bang next to the cemetry wall. Well, imagine you live in one of them, it's a bungalow & your bed is right against a wall thats next to the cemetry wall.

Well, right next to that wall on the other side (these places can get crowded these days) theres a grave with in it a dead body. Well, that dead body may be respectably buried, but youre nearer to it (only a couple of yards) than I am in my bed to a certain cellar. Yet you in the bungalow think nothing of it. So neither do I.

I expect if anybody was to read all that – they wont, by the time I've burnt it! – they would think I was mad. I'm not. *They* might do things *I* thought were mad – people just dont understand other people.

At Office yesterday, a mother brought a child with infected teeth, poor little girl was crying herself sick, with pain & of course frightened of me. I got my D to hold one hand, I held the other, for a minute (it works wonders).

She soon calmed down, & the mother said (which I'd heard before) Doctor what a soft touch you have.

Then my D said, Peter why do you look amused at that? And of course I couldnt tell her.

Re the whole operation in No. 39 – one funny thing. By the end, with all the tidying & washing, as well as using up all the old newspapers for soaking, I had run out of towels, & with contents of hole in cellar ready to be covered in (lime, & clay from garden all ready) I wanted something to use as towel to wash my hands when finished.

So went upstairs to pick out an old shirt from my chest-of-drawers, & what should be the first thing I fish up, but a Pajama-Top!

I had to smile, & nothing could stop me from taking it down to the cellar, looking down into that hole, & then I said out loud, This is from Mr P.T., & dropped it down & then covered the whole thing over with clay & lime, & the bricks over that.

Childish, I know, but everybody does something childish sooner or later.

Which reminds me – must spread coal-dust & bits of coal back on to disturbed part of cellar floor, also that filthy broken chandelier thats been down there since the first night here, when it crashed down. Pretty grand ideas we had in those days, didnt we?

Sat. Feb. 5, 7.30 am

Yesterday eve, as I got in at the gate Mrs. Lucas the char was leaving, I explained that the lady of the house had been called back to U.S. on business, family illness. She said she'd never seen the kitchen so tidy.

Slept like a log. And today lunch, Holborn Rest.

(Rest. for Restaurant, but it does mean *Rest* to me!)

And then Walpole Hotel. I feel Ive deserved it.

Wed. Feb. 10

Today is the day I should have gone to Dieppe . . .

Sun. Feb. 13

God what a long empty day. And my D sitting in those dingy digs in Constantine Rd., Hampstead, it isnt right. Ladies Guild Meeting tomorrow, Albion House, so Martinettis will be popping over to my Office, any news of our dear Belle?

MEMO of Talk I must then have with them etc, Got a cable, her sister pretty sick & she'll be there some time, would I send over more clothes etc.

Mon. Feb. 14

I was right, in they came. Gabble, gabble, we cant *think* why she didnt leave a note for us. Then Mrs. Eugene Stratton looked in, *Do* let us know how she is.

They all seem quite fond of her, I cant understand it. But you never know with these Stage Folk. They didnt even glance at my D at her typewriter. She was relieved.

I asked Dr. Rylance if he did not notice I looked lonely, he remarked that he did.

Tues. Feb. 15

I said to my D this morning, to celebrate our Saturdays, lets have a Night Out tomorrow, early dinner in West End, then Seymour Hicks in *Captain Kidd*

And, I said, I want to show you something. And playful, I take her arm, march her into my private room, & unlock the safe, and take out furs, fur muff, and jewelry: (a) marquise ring with 4 diamonds & ruby (b) gold brooch with points radiating from center set in brilliants, diamond at center.

And more. I look at her, smiling.

Now I'm expecting her to say, Oh Peter, how wonderful of you. No. She just looks kind of bewildered.

Her: But – why didnt she take all this with her to America?

Me: Because I wouldnt let her, thats why!

Her: But when she left, you werent there!

I was ready.

Me: During that row the night before, she kept saying over and over again, I'm leaving you & packing up everything of mine & I'm off *tomorrow*! Not with anything *I* own, I said, not with jewelry or furs – I was in such a rage that I went into that bedroom and picked up the lot and stuffed it into a suitcase and brought it here. Next day, all dressed up to go, she must have shook those empty drawers and as good as spat at them! (I enjoyed the picture of that).

Her: But Peter they were presents, theyre hers!

Me: No my love theyre *mine*, for you to have. With her vulgarity they looked a bit showy on her, but on you they look lovely.

A less nice girl than her might have said, And I suppose when the time comes for the 2 of us to split up it'll all be yours again – but my D wouldnt think of such a thing. All she said was, Oh Peter, I dont know what to say, I'm at sixes & sevens!

Thats the difference.

Wed. Feb. 16

At the Office this morning I brought it up again & said, You'll wear some of your things on our outing tonight, wont you?

The dear girl started off saying no no no, she was quite startled at the idea, but I persuaded her in the end, saying that it was a 1000 to one chance nobody we knew would see us in the West End, then she said no again & finally we compromised on the gold brooch and furs.

Well, we both changed at the Office, she looked lovely in

fox fur with her delicate hands in fox fur muff to match, plus brooch. She was still not sure if it was right but being a woman had to enjoy it a bit.

But after the little dinner, walking down Shaftesbury Ave., what do we run into (that 1000 to one chance!) but those damned Martinettis, also off to a theater. I barely introduced my Secretary, they gave her a funny look – Mrs. M. in particular, I saw her move her face forward as if to have a closer look, & she seemed to give the brooch quite a stare. Then we moved on. Impertinence.

Proceeding to theater, my D was quite taken aback by this incident & it was quite hard to cheer her up. But the play did, so all was well. Seymour Hicks very polished.

But I did hate seeing her off on the omnibus to her Hampstead lodgings. My D, you belong in No. 39.

Thurs. Feb. 17, 8 am

Seeing my D at the theater last night has given me the urge to show her off in public, so must now put together Memo of Talk with her at Office today. My D, I see that on this Sunday Feb 20, the Music-Hall Ladies Guild is holding their Benev. Fund Charity Ball – now I think we deserve a real night out (not just a theater) & I can get tickets through Mrs. Eugene Stratton, ½ a guinea each, and you can wear my favorite Princess robe I gave you with the gold sequins, will match the gold in the brooch on your bodice.

But my D, the brooch is *yours*, from me, & anyway Clara Martinetti had a good look at it & isnt going to keep her trap shut, what does it matter anyway? I *insist*.

Mon. Feb. 21

It was a beautiful ball. Plain sailing.

Well, not *all* plain sailing, she had really jibbed at wearing her brooch with the Martinettis & Nashes etc recognizing it. I said, that is just *why* I want you to wear it, to make me proud of you. And however they pretend to have liked her, she did leave me, didnt she?

Then she said, But Peter *they* dont know that, you wrote she'd gone to America because of family illness, *why* didnt you tell them she's left you, it would look so much better for us now!

And of course she was right – but we all make little mistakes.

Well, at the last minute, as we pulled away from the house (Id told her to get cab and pick me up on way to Ball) she starts to take the brooch off. But I was very strict with her & told her I didnt like to think of her as a coward, & she gave in.

And as to the Ball – I was no Fairy Prince of course but Cinders looked a treat as the Londoners say, every inch a lady, & I loved to watch the admiring looks. A couple of dirty ones as well, but thats life.

Mrs Smythson asked (my D out of earshot at the time) if I had heard lately from Belle, I said, Oh yes, yesterday, shes right up in the wilds of the mountains of California.

Oh good, she says, and how is the patient?

Oh, I said, (stupid) she's not sick.

No, she says, I mean the relative she went over about, who was so ill.

Oh, I said, much much better.

Then Mrs Smythson goes on – Isnt that lovely, means Belle will soon be back with us, she must miss the good times she has with her friends in the show business – I'm sure the next to hear from her will be Clara Martinetti, whos waiting anxiously.

Food for thought. This could only get more & more awkward. Have I got to tell the truth? That she left me for Bruce Miller, and has started a new life in the States and wants to turn her back on anything to do with her life with me, so no communication with her London friends?

Well, when I say the truth – !

Anyway, I dont fancy passing on that information, it being highly detrimental to me as a husband & a man.

Before they start really wondering why they havent heard, I must put my thinking cap on.

Tues. March 1

My goodness, the house is empty! Not empty of what was living in it before – empty of my Dearest.
MEMO of talk coming up, to her. Being a decent girl, she will be shy of the idea & I must be careful how I put it.
My D, the Saturdays are lovely but it is in No. 39 that you belong & I cant rest till youre beside me in that double bed bought by me & which I left years back. Come where you belong my Dearest.
How do you mean it would be a haunted house? Now dont you talk like one of those novelettes, theres nothing of her in the house except all those trashy clothes and gaudery she left behind, & Ive piled them out of sight in one big cupboard.
So we can start afresh. Take your time & think about it. I must gently remention it, every other day.
Yes, but *Ive* got to think about it too. We want to get married, We are *going to get married*. But by the law we cant until desertion has gone on for 7 yrs & the person assumed *dead*.
This calls for thought.

Thurs. March 3

I was right, she's like a frightened bird at the idea of moving in, but every day I make a bit of progress.
MEMO of next talk I must make to her.
My D, you'll be my official housekeeper, what does it matter what people might hint at? All by myself in that big house, I'd simply waste away. Mrs. Lucas only comes twice a week, on those mornings I can ruffle my old bed in the other small bedroom & nobody any the wiser!

Sun. March 6

MEMO of next talk. My D, lets make it next week-end Sat. 12th, lunch at Holborn Rest., but *not* the Walpole Hotel, lets make it No. 39, & for good! And that means we'll wake up, in No. 39, on my birthday!

And in the evening, a quiet home dinner with wine & you can give me a taste of the cooking you boast about – no my D, Im wrong, you never boast – I wouldnt like that.

Sun. March 13

What happiness. My 48th birthday, and she did move in. (Had never seen inside the house, of course.) On the doorstep, Welcome home I said, picked her up & carried her over threshold, I really surprise myself these days.

And a beautiful dinner just as planned, & her a good cook, she'll soon get used to the kitchen. Nervous of course, & in the sitting-room kept looking round at everything, she will soon settle down.

Doesnt mind the pink.

And for the first time in No. 39, I slept in the double bed. A special *special* night.

She has belonged to me now for best part of 3 yrs, & will, for ever.

And she likes her work – likes being a help to me, & I like mine. Folks ask me, dont you find dentistry so finicky, all those gaping mouths? No, I like it *because* it is finicky, & as to the mouths one mouth is much like any other, some with more teeth than others, give & take.

Memo, go out tomorrow & buy her something for *her* birthday. A bangle.

Sun. March 20

A good week of settling in.

Tomorrow early will take cab to my D's old lodgings, 80

Constantine Rd., Hampstead, with suitcase full of clothes no longer corresponding to No. 39, for that nice landlady of hers Mrs Jackson – shoes, long cream coat, mole coat & skirt, black voile blouse, hair-combs, 3 night-gowns almost new, pair pink shoes, black feather boa etc etc. She deserves them, for having been like a mother to my D.

Since Mrs. Lucas is only twice a week, my D will do only 4 days in Office so as to run house etc. In the Office will miss her sorely on her non-Office days, but the thought of her settled in my house will more than make up.

Now plans, I like plans. Some days back, I wrote that *something calls for thought.*

It does. And I have the solution to the problem. Well, 2 problems. And the solution would be (excuse the expression) to kill 2 birds with one stone. That is, it will (a) stop the busybodies wondering why they arent hearing from dear Belle & (b) it will make my future (with my D) that much nearer.

First – Dr. Masters has a married daughter in N.Y. & gets cables from her, Xmas & birthdays etc, & pins them over his desk – last week I saw him put old ones away in a drawer. Slip into his office tomorrow lunchhour & take one from drawer.

Draft letter from me to Martinettis. *Dear Clara & Paul, Please forgive me for not running in during the week but I have been really so upset by very bad news from Belle that I did not feel equal to talking about anything. I know you have been puzzled & worried not hearing a word from our dear Belle, and so have I the last couple of weeks – well the explanation, and not a happy one, is she has been too ill to write, & gradually worse, so much so that I am considering if I had better go over at once. I do not want to worry you with my troubles, but I feel I must explain. I will try & run in during the week & have a chat. With love & best wishes, Peter.*

My D & I had a beautiful Sunday today, more like summer than March, fine & dry so I get to digging at flower-bed in front garden like a countryman, my D has shopped for some bulbs.

MEMO. Talking of summer, *must do something about suitcase in attic* before warm weather comes.

Midnight

Soon after writing that, I heard something in bedroom, & theres my D crying. I put my arms around her & she said she was ever so sorry but just has bouts of not getting used to No. 39, she's tried but just cant.

Then something about the house being weird, no more than that. I told her it was natural, & cheered her up gently.

It has given me a slight upset stomach. And I had been feeling so well.

Mon. March 21

Mrs. Lucas now comes every day except Suns. A real regular household.

Funny I should have written that re suitcase & warm weather – the matter has got more urgent.

Tonight I was complimenting my D on the place being in apple-pie order – the way she's switched things round & bought nice things at Caledonian Market etc. Oh, she said, Peter I am glad, now I can start on the box-room & clear out the rubbish.

I didnt say anything.

MEMO Talk. My D, I dont want you to go up there yet, theres stuff belonging to *somebody* else that shall be nameless which I pushed up there, so you wouldnt be upset seeing it lying round the bedroom etc, wait till Ive got rid of it.

Tues. March 22, 8 am

Ive got a good idea.

MEMO of Talk, which I should slip into v. soon. My Dearest, I tell you whatll help this – here we are with Easter coming up, & a couple of days change of air will do the trick & youll come back fresh to the house, a different girl – you remember my business trip to Dieppe, well I thought then what a nice place to take you too, we'll pack a couple of suitcases, Easter Thurs till Tues – what do you say?

And talking of packing, my D, Ive got an idea – since you want to clear the attic out, I'll travel with that nasty suitcase full of her rubbish – Ive never been able to fancy opening it – & when nobodys looking I'll tip it overboard.

Good riddance.

10 pm

My D was home today, I arrived back at 6 & said I'd had a cable. I took it out of my pocket (one of the Dr. Masters ones) & read out, *Belle dangerously ill with double pleuropneumonia. Otto.* Then put cable back. My D was upset, she has a really kindly nature.

Wed. March 23

This morning, at big entrance into Albion House, ran into Clara Martinetti and Mrs. E. Stratton on way down from Committee Meeting, Ladies Guild.

They button-holed me of course, though could see I was in a hurry. I took the cable from my pocket & read it out to them (or seemed to) explaining *Otto* was my son & that I had a feeling there might be worse news later, in which case I would be going to Dieppe for Easter.

People are strange – they stared at me as if Id said I was going to throw myself off the top of Albion House, & Clara said, Oh Peter, you mean if the news gets bad you'll go, but why? I said, For a change of air. They stared even more, & I said, Of course, I shall be so upset I will have to get away & pull myself together. They said, Oh dear & went. I was polite, as always. People are strange.

Thurs. March 24

Home early from Office, so my D can pack, she's at it now upstairs, excited, we're off to Dieppe tonight. In 10 mins I'll take up a cable (same Dr. Masters cable), this just arrived, then I'll read out to her, *Belle died this evening six o'clock Otto.*

(My D, we cant cancel Dieppe now, it'll help me forget.)

Then send (from Victoria) telegram announcing news to Martinettis. *Belle died yesterday 6 pm. Please telephone Annie Stratton. Shall be away a week. Peter.*

Telephone stage paper *The Era* to have Obit on Sat. ELMORE, March 23, in California, U.S.A., Miss Belle Elmore (Mrs H. H. Crippen).

Mail 1/6 Postal Order to *Era*, to pay for Obit.

N.B. Order blackedged mourning cards, plus writing paper. Also, buy black tie, black suit, ready-made.

Tues. March 29

Back from Easter in Dieppe, lovely holiday. Light suit. For the rest of last Thurs, of course, in Dieppe, my D was upset (over my son's cable) but the sea trip brought roses back to cheeks.

We sat in same cafe I was in before, & I was able to show off my French, she was so impressed she said she must learn some too.

Then we got into conversation with the waitress, a nice plain girl (spoke quite good English, as Dieppe people can do) who told us her dream is to go into service in London (as said) but the type of day-tripper that comes over doesnt offer that sort of chance – anyway she wants to go into a house which her mother would know beforehand was respectable.

Then my D said jokingly, Wouldnt it be nice if she came to work for us & Id get so my French was better than yours – it ought to be, with a name like Le Neve!

Well, I put it to the little Mamselle, & when I told her I was a Doctor she jumped at it. And next day her Maman comes over & she seemed to take to us, saying she would be relieved to know that *ma fille* will be with a Doctor *comme il faut* (respectable).

I take that as a sizeable compliment.

Her name is Valentine, I call her our Easter Valentine. Will come to us a week after we get back, will explain to Mrs. Lucas we need a maid to live in. Showed V. (from my wallet) a snapshot of No. 39.

Then we joked about us going to have a French maid, like in those farces with Charles Hawtrey etc.

(PS – Suitcase disposed of on Channel trip out, Newhaven – Dieppe, no problem. The fishes are welcome. As the French put it, *bon appetit*.)

Lovely holiday.

Wed. March 30, 8 am

Today will be Music-Hall Ladies G. Committee Meeting, so must be ready for the condolences – oh yes, what a shock, cant believe she has gone from us etc.

9 pm

I was right, those very words! Also the Guild want to send an Everlasting Wreath, & I was quick & explained that the term *Music Hall Ladies Guild* would convey nothing to my son (trying to put it as tactfully as poss.) so no point. Anything else in the cable?

No, I said, he just put, was at bedside at end & held her hand. (My son Otto in the old U.S. days couldnt stand her, but it sounded right.)

Oh that is touching, they said, any details of illness?

No, I said, it *was* a cable, & cables cost money. One up to me.

Oh well, says Mrs. Martinetti, itll all be in his letter, you shld be getting that quite soon, we would like to know, having been so attached to her. (I thought they would never go. I wish people would mind their own business.)

I told Dr. Rylance I hadnt let him know before, so as not to spoil his Easter.

Wed. April 6

Black-edged writing paper arrived. Draft letter to her half-sister in Brooklyn:

My dear Louise & Robert, I hardly know how to write to you of my dreadful loss, the shock to me has been so dreadful that I am hardly able to control myself. My poor Cora has gone, I did not even see her at the last. A few weeks ago, we had news that an old relative of mine was dying—

(No good saying it was a relative of *hers* – as I had before – her sister would know it wasnt.)

—and, to secure important property for ourselves, it was necessary for one of us to go. As I was very busy, Cora proposed she should go, and return by way of Brooklyn & she would be able to pay all of you a long visit. Unfortunately, on the way my poor Cora caught a severe cold which has settled on her lungs (etc). She wished not to frighten me, so kept writing not to worry about her.

(Etc – then about the pleuro-pneumonia, then –) *the dreadful news that she had passed away. Imagine the dreadful shock to me – never more to see my Cora alive nor hear her voice again—*

(Wait – I have just had what they call here a *brain wave* – in U.S. called a *brain storm*, funny that) I just remembered that a couple of years back, her mother died, in Calif., & her ashes were sent to No. 39, & how the daughter carried on sobbing about it, wouldnt have the box near her, it's in a corner of the box-room right now. (Where a box belongs!)

So I continue in my letter to Louise:

A small comfort is that her ashes are being sent back to me and I shall soon have what is left of her, here. Of course I am giving up the house – in fact, it drives me mad to be in it without her. As it is so terrible to me to have to write this terrible news, will you please tell the others of our loss. Love to all. (End of letter)

Valentine is a treasure & loves my D. At the Office my partners Dr. Masters & Rylance & even William Long pull my leg about my French maid – they just wont believe she's as respectable as I am!

Last Sat. afternoon, with me having a siesta after busy week (a lot of toothache about, all of a sudden) a ratatat at the front door. Clara Martinetti, if you please, with a taxi waiting with in it Mrs. Eugene Stratton plus nephew. Luckily my D (my 'housekeeper') was upstairs. Just passing, she said, wondered if you had more news from your son – by the way (& this was real prying) what was the name of the boat poor Belle sailed on, from Liverpool? The Steamship *Touraine*, I said (it just came to me like that), and thank you for calling.

House & front garden lovely. And I'm sleeping so well.

Thurs. April 21

My D said at breakfast, it's exactly a month isnt it? (She wasnt going to add the words *since she died*, she would have hated to, I can understand that.) It seems odd your son hasnt written details, dont you think?

This had to come up sooner or later & I was ready for it. Oh, I said, his letter arrived a week ago – to this house & not to the Office, thats why you didnt see it – & I didnt want to depress you with bringing it up. All very sad, he enclosed the Death Certificate. (I added that on an impulse, thinking no-harm-in-that, sounded good.) Then I changed the subject.

But I'll get on to brighter things, it's the future we must look to, & not the gray days that are past. A very nice thing has come to pass – Mr. & Mrs. Masters have asked me to dinner, *with my D.*

I had gone there a couple of years ago (saying my wife wasnt well, because she'd have sure put the kybosh on the party as the Londoners put it) & I liked them & they liked me. Now it is really kind of them, & a snub to the gossips – I know they will both like my D.

Sat. May 7, 11 pm

Nice dinner party, at our home this time, just us and Mr. & Mrs. Masters. My D a delightful hostess, shy & friendly. Party only marred by last nights shocking news of King Edward's death, the end of an Era, albeit a short one.

But it did lead to us getting on to subject of medical etiquette: ie the fact that any doctor gets wind of all sorts of scandals & has to keep them to himself. Well, Mrs. Masters tells me that Dr. M. was for many yrs (before something went wrong & now he has to work hard at ordinary practice) assistant to one of the big Physicians by appointment to the King himself, & then she did hint that her husband has got to know details which would be v. interesting in years to come.

I was taken with that & did coax him to tell a couple of anecdotes. (Here I will mention no Royal names, in view of the present Royal bereavement.)

Very pleasant evening, a glass of wine each.

Thurs. May 19

After 2 weeks, the Nosey Parkers are back.

Who should I run into, in Maples Emporium (Tottenham Court Road) but Mrs. Nash & Mrs. Smythson – me with my D on my arm.

Whilst I was raising my bowler – still in my mourning of course, black tie – they started quite rudely at my D, & I'm afraid she was so timid she just walked on. Then we chat quite stiffly – so sad the King dying like that, poor Queen Alexandras so deaf & so brave – then Mrs Nash says, Did

you get news of the funeral? (In quite a cold tone of voice).
I was at sea for a moment, & said, The Kings funeral? Well they gave me a terrible look but I hadnt said it as a joke. No, they said *another* funeral. Both ladies sticking out bust-wise, & glaring at me from under those big hats, like a couple of haughty hens.

Then I remembered my brainwave about having her mothers ashes, & Oh, I said, the funeral went satisfactorily, all was over very quietly & I have her remains at home.

Well, they just stared, and when I realized what I'd said I could hardly keep back a smile. I hadnt meant it to come out like that – but it did! Then I explained about poor Coras ashes, and said that if interested they would be welcome to see them any time. (I nearly said *examine them*, I just stopped myself.) The Nosey Parkers just bowed and walked off towards a shop assistant. Sanctimonious nobodies.

Our evenings at home get even nicer, more and more cosy. Valentine adores my D & even the cat next door comes in every other evening by the kitchen window to say hello to her & ask for a saucer of milk.

The 2 of us dont seem to be parted for a moment – she said once, Peter, wouldnt it be nice if you sat with me in the kitchen reading the paper while I get our dinner ready.

So now we make quite a habit of it, I even fetch & carry salt pepper etc & watch the oven while *she* has a read. Let me write down how delicate & pretty she looks sitting there, with her brown hair like a sort of halo against the lamplight.

And we talk by the hour, in such an easy manner. Like when she told me her real name is Neave, but when she was 14 she thought it was too ordinary & when she heard there might be French blood in the family she turned into Le Neve.

We tell each other everything. Plain sailing.

Last evening after I had peeled the potatoes (still good at that, like in former days!) I was having my usual read, about the plans for the new King George's Coronation, when my D shuts the oven door on whatever was cooking for our supper, looks thoughtful, and goes out and crosses the passage to the cellar.

Whats on your mind my dear, I called out. Oh she said, very calm and collected, it's the cellar. I want to examine something.

I sat staring at my paper, but the King's head in the photo looked pretty jumpy & blurry & no mistake.

Not a sound. I sat calm as a cucumber but I thought, Im not enjoying this waiting, not a bit.

Then back she comes, holding up a dusty tangle of glass. Peter dear, she said, this must have been so handsome, couldnt we have it repaired? The smashed chandelier.

Oh no my dear I said, I do appreciate your thought but the things too far gone, just leave those pieces with the rest of the cellar rubbish.

Which she did, and by the time she came back I was at my newspaper again. Then I looked in front of me, thinking, and I must have looked pretty thoughtful to make her say then, as she was laying the table, Peter are you finding this life a bit dull?

That was unexpected. I said, My Darling what do you mean, Im as happy as the day is long!

I know, she said, but you look sometimes as if you wish something would happen. I asked what on earth – what sort of thing?

Oh, she said, I dont know, just my imagination – something exciting!

Looking back into my paper, I said to myself that it was downright silly, but she *is* some sort of thought-reader. Because though I am a most contented man, there are moments when I feel an urge to have the jog-trot (as it were) of the daily round to be interrupted by—

By what? Well, to start with, by me able to prove to people – now I dont want to sound swollen-headed even on paper, but it would be nice to show all the ordinary people who take me for granted, how clever Ive been.

And brave too, not many brawny heroes wearing the V.C. from the Boer War would have the nerve to go through what I was forced to, and succeed. And nothing to show for it, whats the good of a V.C. if youve pinned it on your chest yourself, & it's invisible into the bargain?

I know thats nonsense, imagine me blurting the whole thing out to my D just to show off, she'd faint dead away & I'd never see her again, no no. And yet – I suppose its only human nature, to want your cake and eat it. There *is* something dull about life now, no excitement – but what can I do?

I must just look upon the plain sailing as what I planned for, & what I deserve. But it's good for me to ponder on private things, in this way.

Wed. May 25

Funny I should have written *plain sailing* a week ago, now I think of this evening. Not that it was a storm, just a breeze really.

I had a little extra desk work & was a bit late getting to her in the kitchen as we had arranged, & found her sitting at the table kind of knotting her hands & then unknotting them.

I said whats the matter, she said she didnt know what was making her like this. I said, I tell you what this Sat. we'll have a night-out West-End style, what do you say to *Dr. Jekyll & Mr. Hyde*, with H. B. Irving?

Oh no, she said quite sharp, something cheerful. I said what about Gertie Millar at the Gaiety, *Our Miss Gibbs*, & dinner before at the New Gaiety Rest. next door, you can wear the Princess robe etc?

Then she said – it wasnt a row, good Heavens no, never could be a row with her, she wasnt even raising her voice, but begging me, like a child thats worried.

She begged me that if we did go out Sat night, I would let her *not* wear the jewelry or the fox furs.

Well, I was quite took aback & said, why on earth not, theyre *yours*, with my love!

I know, she said – I know, but – Peter how can I explain – I have a feeling – that Im being watched.

I stared at her, as was natural. By who? I said.

She said I suppose its because she's dead, but every time I put them on, I can feel her eyes on me, sometimes in front &

sometimes behind & thats the worst, makes me feel quite ill—

Then she started to sob, & of course I took her in my arms like she was 10 yrs old, & to see her cry made me feel ill as it had done before. Then I give her a drop of brandy & she calms down. Then I did a lot of persuading & she realized Id be upset if she persisted with this & that it was weakness on her part, & she gave in & will wear her presents.

She not only loves me, she respects me. Women do get superstitious & I must humor the poor pet. Things take time.

Tues. June 28

It was such a lovely evening, after day in Office my D & I sat in our bit of back garden, near the roses, you could smell them. She has green fingers.

Then a knock at the front door, could just hear it. Mr. Nash if you please. Bluff, friendly, but a bit – whats the word—

Yes, *furtive*.

The impertinence of some people. Just a social call – then he says, Have you heard any further from California?

I said, No, not since I had a letter from my son, after the funeral. Then he said (I could hardly believe my ears) Have you got the letter here?

I said No, its at the Office.

Then I took pleasure in playing my strong card. But, I said, I'll tell you what *is* here in the house – Mrs. Nash must have told you, with her being so interested. My poor wife's ashes. Theyre down in my study.

(Then I quickly had to turn to my D & explain that the said ashes had arrived on a morning when she was out shopping before we both left for the Office, and that I had put them away without mentioning them to her, so as not to upset her with such a gruesome matter.)

Well, I then said to Nash, very polite I was – would you like to see the ashes?

He goes quite red – no thank you Peter. Then he said, what his wife & Clara Martinetti hadnt been able to understand, was me going off to Dieppe after hearing of the decease (oh these women!). I said I had told the 2 of them, *for that very reason*, to try & get over the shock. (He didnt mention my D having accompanied me to Dieppe.)

He went quite soon after. What I didnt like, was that he had quite spoilt the evening for my D.

Impertinence. What news from Calif. indeed, what did he expect when somebody's been dead for 3 months – Spirit Messages?

I just want to be left in peace by such people.

Tues. July 5

V. sad evening for both of us. Big upheavel at Office.

Poor Dr Masters (always delicate) had been home with the flu for over a week, then it turns to pneumonia. A daughter telephoned the Office in tears, he died early this morning.

Such a nice man to have gone so suddenly, so cruel. My D wept into her hanky which made me feel worse, I had to pat her hand.

Funeral on Sat. the 9th, I told her we'll shut the office for whole morning & go to pay our respects. Such a devoted famiy.

Wed. July 6, 9 pm

Well well well—

Feel all stirred up again, must write it all down—

Today was my D's day at No. 39 going to wash curtains while Valentine the maid cleaned out basement. In the Office, I had just finished work on a lady patient (a stopping) & thinking of my D at home, hanging curtains on clothes line in back garden – well, the patient was hardly out of the door when there was quite a scurry of feet on the stairs & the Office door burst open.

It was my D, quite out of breath. I jumped up and said, What is it?

Scotland Yard, she said.

PART TWO

The Yard Versus Crippen

8

JUST ROUTINE

Got to write it all down fresh, like those statements criminals make, so I can examine and sort out.

When she said, Scotland Yard, my mind flashes back – how the mind does dart about, when called upon – I hadnt heard *Scotland Yard* (except written down, as in E. Wallace) since I was a young chap come here from U.S. to watch operations etc, when the policeman remarked that it wasnt a Yard & not in Scotland.

As she said it, behind her I see two bowler hats coming up the stairs.

Then the two gentlemen in the doorway, one my age with clipped moustache, the other a young fellow. Says the older one (v. pleasant) Excuse the intrusion Doctor, I am Chief Inspector Dew, Metropolitan Police Office, New Scotland Yard. Tall, baldish, good trim figure of a man, features sort of gaunt & spare but a pleasant expression. Quiet, quietly dressed – not quite a Gent of course, but well spoken. Then he said, This is Sergeant Mitchell. A well-behaved Cockney lad.

(My D told me afterwards that at 10 am she had been upstairs taking down curtains when V. comes up to say 2 gentlemen to see her. My D explained to them she was the housekeeper, the Insp. told her they had called to make inquiries about a Mrs. Crippen.

I asked her if they had the cheek to examine my desk or start wandering around, she said no, they just sat with the Insp. chatting to her. She did notice the Insp. had a good look at the little brooch she was wearing.

Then the 3 of them came down to the Office on the omnibus. And here we are.)

Insp. Dew says, Doctor could we see you alone?

(It was nice that the Insp. called me by my title.)

My D looked quite flustered, but I pressed her arm & said, Im sure its nothing. I show the gentlemen into my private room and ask them to sit down.

And gentlemen they were too, in behavior. When you think of the cops in the States, Ive never met any close to, but just to look at them, hulking mannerless louts. These 2, in their bowlers & stiff collars, might have been 2 other Dentists come to ask me if I was qualified to adress a Conference or some such.

I wasnt scared. And not scared now, I just want to straighten out. The Insp. takes pipe from waistcoat pocket.

D*ew*: Mind my Navy Mixture, Doctor?

(I said I didnt, he lit up)

Dew: Last Thursday a Mr. & Mrs. Nash called on us at the Yard. Just a chat.

(*I was right* – I guessed theyd talk, but never thought theyd have the sauce to go as far as this)

Me: Yes Inspector, what did they say?

Dew: I'll read you their statement.

And he did. We wish to state that we have been v. worried about our friend Mrs. Crippen disappearing as she did, etc. Then about noticing the brooch on my D – a lot of busybody mountain-out-of-molehill stuff.

Cant recall details, but one bit struck me as funny. Mrs. Nash had said something like – We cant believe that Belle wouldnt have written us her adress – & if she did, & then didnt hear from us, shed have been quite cut up.

Cut up? It was quite an effort not to smile.

Dew: This is just a routine inquiry of course, & we thought you could help us to ease their minds plus the Martinettis, they being close friends of your wife.

Me: Theyre all nice people Inspector, but Im afraid they love a bit of gossip & if they can make the gossip a bit spicier than need be, theyre inclined to do so—

Dew: Quite quite. At the Yard, we have to deal with a lot of such well-meaning people, for our sins. And now, if you dont mind, a Statement from you. And the usual preamble – just routine you understand – if you wouldnt mind writing down, *I desire to make a voluntary statement to clear the whole matter up*.

1910 JULY 6: HILLDROP CRESCENT

Then he said, Would you mind repeating after me, so Sergeant Mitchell can take it down as being the beginning of your Statement – (then something like) – After having been cautioned that anything I say may be used in evidence, I wish to make the following etc. Just routine.
First, of course, the account of the death. I thought it best to describe my son's 2 cables & his letter after the funeral.
Dew: Yes, Miss Le Neve told me about them. (So my instinct had been right).
Dew: And now, if you dont mind, the details of what one would call your *particulars*. Just routine.
So I talk, with Serg. Mitchell taking it down in shorthand.
From the beginning, born in Coldwater etc etc, the Story of my Life – (Having kept this Diary helped me to remember a lot of the details) After a ¼ of an hour, a knock at the door – William Long my dental mechanic announcing a patient for an extraction. My D had gone to help in Dr. Masters office.
I said I was busy, but the Inspec. said, No no, I dont want to interrupt your days work. So we stopped & I got on with the extraction, later on it was lancing a gumboil, then another extraction, with in between 2 visitors hearing my Life Story. (It was a specially busy day with poor Dr. Masters patients to see to as well.)
They seemed really interested in the work, & I was quite pleased to be seen at it. I felt the Insp. respected me for my job as I respected him for his.
We hadnt got halfway through when it was One pm & the Inspector suggested lunch, in the corridor I put my head in at Dr. Masters to call to my D. we were out for lunch & would be back, they took me to a little Italian place I knew round the corner, nice roast-beef.
It was kind of him, but at same time I knew he didnt want me out of his sight in case I got together with my D, & made up a story about California.
At the lunch-table (now that the notebook was put away) we chatted – really chatted. About the Coronation next June (George V) – my D and I must get into the sidewalk crowd for that – then the Archer-Shee Case (the boy thats supposed to have stolen the 5/- postal order) then about how Scotland Yard works, the files they keep, the finger-print

collection, how they have to go to Police College first & the tough Detective training, wrestling & all that, it was very interesting to the layman as it were.

The Inspector looked like somebody sitting in his club, with his pipe puffing away & eyes twinkling. He told me he had a son same age as Otto, the son studying to be a doctor. I didn't tell him I hadnt seen my son since I came to England, I thought he might think it kind of unfeeling.

He told me he was something of a gardener & I told him about me digging the flower-bed so as to get the bulbs in for Easter. Then – kind of out of the blue – he shot out, quiet & pleasant—

Dew: For Easter you and Miss Le Neve went to Dieppe?

Well I didn't expect *that*. But no good saying no, he must have got it out of the Nashes.

Then I thought, in a flash – that *suitcase* I took away on the trip & never brought back – could Valentine have told them – no, V. didnt see us off, we hadnt even met her yet—

Then, re me & my D, I saw my way clear, & went to work on the Insp. And looking back, I think I did a good job.

Me: Inspector, I can see youre a man of the world & I'll speak freely in front of Serg. Mitchell. My wife & I had not shared a bedroom for 5 yrs, & you know what that could do to a man in good health.

(That gave him quite a twinkle of the eye.)

Me: As you could both see this morning, my Secretary and Housekeeper is an attractive young person, a lady, & with me not getting on with my wife we naturally formed a close bond. Now that I am free she is my fiancee & we plan eventually to marry.

Dew: I can understand that, she is such a charming young lady she deserves to be happy – if I had a daughter Id want her to be like Miss Le Neve.

Now that pleased me. And the little confession had been a wise move on my part, cleared the air.

Well, back to Office & on with my Life Story – my tactful D kept out the room – ending at last with the Martinettis coming to dinner Jan 31. (He must have seen the Martinettis as well, theyd told him she was rude to me at dinner over him going to bathroom etc.)

Then a tricky bit.

Dew: The young lady mentioned at your house that she had an idea Mrs. Crippen had been suddenly called back to the States due to family illness. Could you enlarge on that? I knew my D would want me to *tell the Truth*. Well, the truth as she knew it. Here goes.

Me: Inspector, I have a further confession, & if I make a statement without that confession, it would be perjury, isn't that right?

Dew: Near enough.

Me: Well, here I go. There was nobody sick over in the States, my wife walked out on me. And I'm 90% certain it was to join a lover in California, a variety singer by the name of Bruce Miller, all her friends here had known him. Do you wonder, Inspector, that I wasnt up to being honest about that? Since I knew they would all be laughing at me behind my back? Can you understand that?

Dew: Very well, a natural reaction in a husband, I'm glad youve told me that.

So far so good. Then an even trickier neck of the woods.

Dew: I know this was over 5 months back, but I'd like you to try & recall – She left for America on Feb 1st. When was your last sight of her?

Me: The night before, between 2 & 3 am. She slammed her door in my face & – I imagine – went to bed. I went to bed myself but so upset I couldnt sleep. Towards morning I heard her banging about, I thought she was rampaging as usual but later, of course, I realized she may be been packing.

Dew: You got up at—?

Me: My usual, 7 to 7.30, made my breakfast in the kitchen as usual, & left for this Office at my usual 8.30.

Dew: I am told you seemed a kind & attentive husband – no question of taking a cup of coffee up to Mrs Crippen?

(A twinkle in his eye, but at this point I felt releived that my D was out of the room.)

Me: No question at all, that had stopped some time before. By her own wish.

Dew: And in the evening?

Me: I got home some time between 6 & 7.

Dew: And found the bird had flown?

Me: Quite right. (For an Atlantic flight this particular bird would be overweight.)

Dew: How could you tell she'd gone?
Me: Well, her bedroom door open and stuff all over the place. She was a very impulsive woman.
Dew: Had she left a note?
Me: No.
Dew: No? (As I had feared he would, he sounded surprised.)
Me: Not a line, she wasnt one to apologize for anything. And I didnt need a note to explain that she'd walked out on me.
Dew: Quite so. But since there was no note, how could you tell that it was America she was bound for?

I was ready for that.

Me: Each bust-up we had had, she had stated definitely that the next time, she would leave me & go to the States.
Dew: To stay with her half-sister Mrs Robert Mills in Brooklyn?

I thought, my goodness, that chat of his with those know-all Nashes cant have been a short one. I had to think quick, she was fond of the half-sister and might easily have gone to her. That is, if she'd gone to the States at all, but I mustnt think of those lines, *she did go*.

Then, again out of the blue—

Dew: What luggage had she taken?
Me: As far as I could tell that evening when I tried to check – she had a lot of cases & boxes – she had taken a suitcase with clothes for a journey, shoes etc for a long trip, plus toilet articles of course.
Dew: Of course. Plus her valuables. What about jewelry & furs?
Me: What jewelry & furs?

Which, come to think of it, was a slip, what other jewelry & furs could there be? (Those Nosey Nashes again. I could tell by his tone that the 2 of them had well pressed that point.) I should have said straight away what I said next – that I had given them to my fiancee.

Then I repeated what I had told my D – that in the row that last night she had said she would leave me & that I'd locked the stuff in the office safe because it was my property.

When I said it was mine, he looked puzzled, just as my D had, I cant understand why people wouldnt be able to see that. Then—
Dew: Must have been quite a big suitcase.
Me: A good size, if it was the one I think it was.
Dew: Is there a cabstand near your house?
Me: Yes, round the corner in York Rd.
Dew: You must have known it well, over the years – did you go along and inquire whether your wife had come along during that day with a suitcase, and taken a cab, & where she had taken it to?
Me: No I didnt.
I couldnt say I had, the cabbies would remember that I hadnt. You cant think of everything. Besides, I was pressed for time.
Then a brainwave.
Me: It had occurred to me to go to the cabstand, of course, then I thought – to go & question a lot of cabbies who know me – which of you drove my wife when she deserted me – no, I couldn't have faced it.
Dew: That's understandable.
And he meant it. I followed up my advantage.
Me: By the same token, Inspector, that evening when I found she'd walked out, I just couldnt have gone round the district to call on the milkman breadman et cetera, to ask if anybody had seen her around the house that morning. Could I?
Dew: Of course not. You employ a servant?
Me: A Mrs. Lucas. But at that time she only came twice a week. It was a pretty irregular establishment then.
Dew: Did she work Tuesdays?
Me: No, Mondays & Fridays. (Bit of luck.) On the Friday I told her what I told everybody, out of pride – that my wife had been called to the States through family illness.
Then again out of the blue—
Dew: What about steamships?
Me: Steamships?
Dew: There would have been no danger of anybody knowing you in any Steamship Office – didnt you think to

ask about sailings to New York that day or that week, to try and trace a woman answering the description of your wife?

Me: I did not, & Inspector I'll tell you why. I was so sick of my marriage that I felt nothing but relief at the thought she was out of my house for good. Why should I want her back? (This was so true that the Insp., whatever he was asking in the line of duty, must have been 100% convinced of my sincerity.)

Dew: When you told your fiancee that the presents to your wife were your property – did she accept the explanation? (Sarcastic almost.)

Me: I persuaded her to see that my way of looking at it is right.

(And it is).

Dew: Well Doctor, thatll be all, youve been most helpful.

Then the Serg. read the Statement back (a slow business) then I had to write down, *This statement has been read over to me. It is quite correct, without any promise or threat having been held out to me.*

It was 5 pm before the notebook was put away, then some chat. I was just answering a query from the Insp. about is there much connection between ear & mouth.

Then I offered the Insp. a cigar & we chatted again. Then, just as sudden as the times before – he remarked, casual like, but again I wasnt ready for it—

Dew: Could I trouble you & your fiancee to join us in a cab for a short visit to your residence?

Well, I was quite taken aback. Theres a British saying, An Englishman's Home is his Castle, & I'm as good as English by now. But I suppose he has to do his job. I said, very polite, I invite you to look round the house & do whatever you like.

Then I called my D in. She looked a bit keyed up & I was pleased to see the Insp. get up & give her a really nice smile, I felt he was thinking of her as his daughter, & the smile made her easier at once.

We walked to the cab-rank and went up to No. 39, chatting easy all the way. About dentistry, false teeth being mislaid – Dentists always raise a smile just being Dentists, but they can get their own back with a couple of stories of their own.

Well, once home I gave them a glass of sherry then (leaving my D in the main sitting-room, not looking too happy) I started them off on the Grand Tour. I wasnt too happy myself showing them round the big bedroom, where the Insp. asks to see the jewelry & furs. I got them out of my D's cupboard & he duly examined them. Then I unlocked the second big cupboard & showed him the pile of fancy clothes, clothes & ostrich feathers, etc, all stuff my D had never seen & naturally didnt want to see. Then he said, Dont bother to come with us for the rest of the look-round, we'll manage.

First thing I do once they are out of the room and upstairs, is to nip down to my study, unlock my desk, whisk out the locked strong-box (containing this Diary), hurry back to the bedroom and hide the strong-box down among the ostrich feathers etc which theres now no danger of them being interested in. Down comes the Insp. again, and I take him down into the study with the remark that I've unlocked the desk for him to examine. I give myself marks for this.

Then back to join my D in the sitting-room, plus the Insp. and there the 3 of us partake in a glass of sherry. The Sergeant still on his rounds.

Dew: Your very good health Doctor – now about the 2 cables from your son, & his subsequent letter – could I see them?

Me: I'm afraid theyre not here, theyre in the safe at the Office.

Dew: Ah well, another time.

(I didnt altogether take to the sound of that.)

Dew: Now, if we could only track down the doctor in California who signed the Death Certificate, we could get from him the details of the funeral—

My D: Oh, but when Dr. Crippen's son wrote him after the funeral, he enclosed the Death Certificate – didnt he Peter? (My good girl has too good a memory.)

Dew: Splendid.

Me: But Inspector, there is a small drawer in my desk here, with something in it which will interest you more than the Death Certificate! (To my D) Stay here my dear—

He & I go downstairs, enter my study, and theres the desk. I take out a small oak box & plonk it in front of him.

Me: It would have upset her to see this, you see, in this box are my wife's ashes, my son sent them on immediately, it upset me a good deal when they arrived, I dont like that sort of morbid carry-on, & I pushed the box away where my fiancee couldnt see it & so get even more upset than me.

Dew: May I open it?

And he did. There they were, her mother's ashes. And printed in small letters. Burbank Funeral Co. That was a stroke of luck, her mother having died in the same part of the world.

Just a small neat pile, like the Insp.'s cigar-ash. I couldnt help thinking – My, those crematoriums do it thoroughly. They make a difficult job easy.

He shut the lid, quite reverant, & said, Very sad, & I took the box and put it back in the drawer.

Then up to the sitting-room to join my D again. Then the Serg. comes up the stairs & asks if there is a lamp or something for him to be able to see in the cellar.

The cellar.

My D has just stopped at my study door to say suppers ready. I'll continue later – my wrist is stiff anyway, how guys like Mark Twain & Dickens got through those books I'll never know.

11 pm

The cellar, the Serg. said. He wanted to see the cellar.

I told him the lantern was in a cupboard in the kitchen, by the sink, with a candle in it, & I handed him a box of matches. They want down to the basement, together.

The cellar? says my D, my goodness arent they being thorough, not worth the journey, just to have a peek at that poor old chandelier!

Her attitude, quite amused, was of course a releif to me – after being uncomfortable about them going into something so private as our bedroom & handling the jewelry, furs etc.

At the same time I didnt fancy sitting in her company waiting for them to come up again, so I excused myself on the grounds I must go down to the study & lock the desk, which I did.

Then I sat down in the study, and waited. Another 12 minutes by my watch, but it seemed longer. Then, Im glad to say, I got further proof of what a cool customer I am. Staring at the open door, I made myself imagine the Insp.'s bulk coming into view – or more likely the Serg., sent up on the errand, & saying, Excuse me Doctor, have you such a thing as a crowbar, or a pick-axe?

In the dead quiet – as quiet as that long night – in my mind I could hear the question, distinctly.

And yet inside me, no reaction at all. Not hot, not cold, heart dead normal. And I just knew, looking at the doorway & waiting, that my face hadnt got that staring look people get when theyre all strung up about something. No I was just – looking & waiting.

I bend my head & look down at my hands. One on each knee, not gripping at all, just loose.

It did seem a long wait, though.

Footsteps, quick ones, & there they are, both of them.

Says the Inspector, What a sad sight, Doctor.

I just look at him, What was that Inspector?

That cellar floor, he said, what a place to end up!

I felt my hands tighten over my knees. Hard. Careful now. Good thing my D's not in the room.

Then he goes on, Was it an accident, or had you just gone off the idea?

I just go on looking at him. Only thing I can think of to say, is – what idea?

Having a chandelier, he says. Must have looked good.

It did, I said, very handsome. I bought it as a house-warming present, for my wife. It just fell down.

Says the Inspector, Very sad.

Then up to the sitting-room again, a farewell glass of sherry with my D. Then the Insp.—

Dew: Oh, by the way – you said your son's letter & the

cables were here – could I have a squint at them?

I go to the desk, but cant find them anywhere.

Me: I *am* stupid, I just remembered they are at the Office after all, in the safe, like I told you.

Dew: Never mind—

Then he gets up & thanks me for giving him so much of my precious time. Then a joke about the next time he gets toothache, he'll come calling on me again.

They shake hands with my D. (The Insp. giving her the same nice smile, & saying, Goodbye my dear). I saw them to front door, they left. About 8.

I went straight down to basement, opened cellar door & lit a match.

Nothing much disturbed, the old chandelier moved over in case there was something in small pile of coal under it. What they call a routine check.

Walking up to sitting-room, it occurrs to me that theres a fairly bookish word for the Insp.'s whole demenour right through this visit – cant think of it though, itll come to me.

But I am pleased to say I felt calm & satisfied. When I think how jumpy I got when the Bobby called to interview me about that mess the Drouet Institute got into – well, times have changed.

I've got some confidence in me.

(Mind you, I did find myself feeling in trouser-pocket for keys, which include key of strong-box (Diary) & of Safe in Office.)

My D says it was the longest day she'd ever spent.

What she was most anxious about, of course, was her own positon & how far it could embarass me – I think she had an idea, poor innocent pet, that we could get into trouble with the Police for not being married yet.

Then I laughed at her & said about the Inspect. being a Man of the World, then I explained it had just been a routine call, & quite natural after those meddling Nashes & Martinettis sticking their finger in the pie.

Then I added that those people had made a suggestion, which made Scotland Yard want to search around for a letter in the ladys handwriting giving the idea she might be going to commit suicide, which she never would do.

That seems to satisfy her, then she says, He's a nice gentleman – when he said *Scotland Yard* I quite expected to be afraid but he has a kind face. (Thats good.)

Though she did say, I would have thought he'd have taken your word about the cremation, once youd told him about your sons letter. I said its the Routine business again, they must get sick of dotting all the I's. She seemed to see the point.

Then she changes again. While we were in the kitchen after they'd gone (almost dark by now) me reading the *Evening News* & her laying the table, the doorknocker over us gives a big ratatat, & she gives a sort of gasp, gone quite pale.

I go up, its 2 women in bonnets collecting door-to-door for the Salvation Army. (These calls are a devilish nuisance but I have always admired the S.A. for the work they do for the poor & I gave them 6d.)

Well, when I came down again she was standing there, very tense, It wasnt him back again?

Me: Of course it wasnt, he's got the truth now & I dont think he'll bother us again. (As I said it, I almost said, I'll miss his company, because in a way it was what I felt. I liked him calling her *my dear*, & *me Doctor*.) At the same time—

Inscrutable – thats the word I was looking for. Right through, he was inscrutable.

2 am

Can't sleep.

But on the other hand, he could be putting that on.

You might say that these Scotland Yard chaps are v. well behaved. On the easy-going side, I dont see them getting the upper hand of these vilent cases you read about.

Must break off to look – Yes, I just peeked into our bedroom and she's asleep thank goodness, been a long day, & a strain & I think I finally set her mind at rest.

And me? Now lets see – My hand is dead steady as I write,

so I'm not in any way upset. Just a professional man confiding in his personal Diary.

But my heart is beating. I can hear it.

Let me work out why. And I think I've got it.

My D remarked, didnt she, that she felt life was oftentimes a bit dull for me? Well, I reckon it was.

And it isnt dull any more. Ive got to sharpen my wits to such a pitch that I may cut myself with them. True, Inspector Dew & his right-hand man Sergeant Whats-his-Name dont seem the cream of the Police Force in the area of brains – a bit on the slow side – but I cant afford to let the grass grow.

Got to keep one step ahead. The Insp cant think of Martinettis & Co as anything but alarmists & time-wasters, and Ive got to play that up 100%. I've a cool head, so it isnt difficult.

Just tricky enough to give a sharp interest to life. A zest, thats the word, a zest.

Now the letter I've got to have, from my son.

MEMO – yank old discarded typewriter from back of attic. Now make draft of letter to be typed on it—

1473A Jones, San Francisco (sounds genuine alright), Fri March 25 1910.
Dear Dad,
   Here is a letter as opposed to usual Xmas Greeting, but you will have had my sad cable by now telling you of my stepmother's death (they got into touch with me as next-of-kin) 2 days ago (23rd) from double pleuro-pneumonia.
   I'm afraid it's the climate here if youre unused to it, cold evenings on top of hot days. I am glad to say she did not suffer too much, mostly unconscious. Death Certificate enclosed. (Cremation, the ashes will be sent to you).
   I am very sorry to have to write this. All well with me as I hope you are. Moving to Denver next week.
                 Your respectful son,
      Then copy signature (from old Xmas card),
                 Otto H. Crippen.

(NB Remember not to give his real address, 1427 N. Hoover St, Rural Delivery, L.A.)

In morning, must leave home early, shop at stationers not too near house for thin wodge of plain typing paper, return house, ask my D to telephone Office to say I'll be ¼ hour late through tidying up some business at home, fetch old machine from box-room, type letter, take it in pocket to Office.

But where are the 2 cables? Search me. And the Death Certificate. I wish my D had never mentioned it. I shld have left it in California where it belongs.

MEMO – call at Camden Town Public Library & look up copies of newspaper for Feb. 1st on, for names of Transatlantic Liners. No harm.

Thurs. July 7, 7 am

Couldnt sleep, sat up reading a Sherlock H. take in an old *Strand Mag.* I was struck with this one, in it a chap who has (by all accounts) a v. nasty wife, gets rid of her in quite a clever manner – & straight away the author labels him as a criminal.

I suppose because the last couple of days have brought certain matters back to my attention, I dont see the Authors point of view & it annoys me.

Ive never diddled a soul out of a penny, all my life Ive been a model citizen, *I am no criminal.* And I dont see that the unusual event of this year turns me into one. I feel quite touchy about this.

Come to think of it, its funny too that while Ive never got any special pleasure out of being the Model Citizen mentioned above, all these years, I do feel personal satisfaction with having achieved my independence through my own unaided enterprise in dealing in a bold manner with a very tight spot, & able to say (like in the Army) *The manouvers were 100% successful.* Im well aware Im in the minority in this, & the great thing is that my D must never know. The idea of her being hurt or worried over me – no, it mustnt happen.

MEMO for the journey I'll make later today, home from Office on my usual omnibus. I'll be carrying my brief-case – who's to know that on this trip it's empty? As I get near home, open it and at destination leave it there, on my seat, open.

MEMO Talk to my D on getting home—

I've just done a stupid absentminded thing. Left a briefcase on the omnibus.

(Then she's bound to ask if there was anything important in it).

No my dear, just some prescriptions & – well, nothing important to me or to you, though the Inspector with his thorough methods would think different. I mean theres the letter from my son enclosing the Death Certificate, I remembered I had it in the Office and put it in the brief-case to bring home for safety. Oh wait a minute, heres the letter in my pocket! (And out of my pocket itll come). I remember, what I put in my brief-case with the Death Cert, was the two cables – the ones I read out to you, remember?

Then I say – well, my dear, for the Inspector's sake we must try and track the Certificate & the cables, we'll try the Lost Property, & only hope that if the brief-case turns up, the contents wont have been thrown away out of spite & itll be empty. Neat.

Yesterday at the Office everybody must have been curious about the 2 strangers shut up with me for so long.

MEMO In case they call again – unlikely – I must happen to say to Miss Curnow of Munyon's Remedies (shes a talker) that they were making inquiries concerning the burglaries round about Hilldrop Crescent.

8 pm

I was right about them being curious. This morning Dr. Rylance said, who were your patients yesterday in the bowler hats? I told him the truth, that they were from Scotland Yard. But that they had come to see if my wife had any estates in England she should be paying taxes on.

All according to plan – this evening arrived home about 6.30 without brief-case, I explained it all to my D. She said she'd go to the Lost Properties tomorrow & try to track it down.

Then I go into study, Valentine knocks & comes in. She looked somewhat sureptitious.

Val: (in her funny English) Two gentlemen today.
Me: The ones that called yesterday?
Val: No. I look out of window to see if it starts to rain, and I see them.
Me: Where?
Val: In garden.
Me: What were they doing?
Val: They look.
Me: What at?
Val: The round part of garden?
Me: You mean the flowerbed.
Val: Oui. Pushing the ground with their feet, then they bend & look close, I think for a minute they will pick flowers & I think I must go out & say to them that this is private garden, & I go out.
Me: What did they say to you?
Val: That they are Police, then they go.
Me: Merci Valentine, itll be those burglaries down the road, I expect they were looking for footprints in the garden. Dont tell Miss Le Neve, these things make her nervous.

I go back to the sitting-room, a knock at front door.

The Bowler Hat Bros, Inspector & Sergeant. Interview yesterday v. satisfactory, just a couple of routine questions. Im getting used to that word routine.

When I took them up into the sitting-room I noticed my D seemed much better after me soothing her down last evening. She had changed into her gray shadow-stripe costume, quiet & plain, I am always proud of her quiet ways.

I wasnt a bit put out by them arriving, if anything quite pleased, for I told the Insp. the whole story about the mislaid brief-case, & *handed over my son's letter.*

The Insp. read it carefully, then said, I *am* sorry, its very sad.

I'm glad, Inspector, I said, to be able to produce his letter for you, as the evidence you want!

Looking back, I dont like the sound of one word – evidence – sounded like we were a police court. But it was out before I could think. He just sat looking at the letter, then put it in his pocket.

Then I asked him if he'd like my fiancee to leave us – no no he said, I wouldnt want to turn Miss Le Neve out of her own sitting-room – & that was a *very* nice way to put it. *Even after what was to happen in a minute*, you have to like the man – behind that smile & polite voice, theres a brain ticking. Tick-tock. Ive got a brain as well. He & me, we're a couple of clocks.

Well, we sat down, I poured sherry & we chatted, about whether Mrs Keppel and the old King's other friends would be at the Coronation next year, etc, then the weather.

It was a balmy summer evening, trees green & shining in the sun. And the room spotless, once again I was proud of my D.

Id almost forgot what he had come about – now I must get this all correct – when he fired a question at me – not fired, no, too strong, it was like the one about Dieppe, kind of casual but sudden. He said, Could you help me, Doctor, about the Death Certificate?

As I was thinking what to say, I thought, he hasnt asked for him & me to be alone, & I didnt like my D sitting there, not at all. Then I thought, if I suggest her leaving it might look as if I was ashamed for her to hear something or other, or – worse – that she knew something she shouldnt be knowing. So I just said – The Death Certificate, yes?

Then, v. carefully, he took my son's letter out of his pocket. I dont see why it's so important.

Dew: Would you oblige us by writing to your son asking for confirmation that he sent it?

It seems to me that was a fairly bold request, but I had my answer ready.

Me: He wont be at that adress.
Dew: Oh?
Me: As you read in the letter, by now he'll be in Denver.

My son's pretty restless, always changing jobs – when he last wrote, he was in Los Angeles, having moved there from San Francisco. It'd be like looking for a needle in a haystack!

Dew: When he last wrote, you said?

Me: Thats right.

Dew: But Doctor he says here – I quote – *as opposed to the usual Xmas greeting*. And that seems to imply he was *not* in the habit of writing you letters?

I had to think, quick, and then be just as quiet-spoken as him. No bluster, Crippen, no bluster.

Me: Im sorry Inspector, I didnt say anything about a letter, I said – when he last wrote, which was last Christmas. On the back of his Christmas card.

(At this point I felt quite pleased with myself, though I was careful not to show it.)

Dew: Would you have the card handy?

Me: Im afraid not. You see Inspector, being a man of method I throw my Christmas cards away on the 7th of Jan., a habit I learnt here.

Dew: (a smile) My wife does the same, says they start cluttering the place up.

Then he said, very crisp—

Dew: Any possibility of anybody else close to her or you being concerned with the funeral?

Me: Indeed yes, who else but this Bruce Miller she went out there for?

The Insp then thanked me, then some more chat, he seemed to pay particular attention to my D, asking her about her cooking, he's a bit of an amateur cook himself, of a Sunday.

Then he got up, brisk-like. You'd never have though he had another card up his sleeve. An ace.

Dew: Thank you again, Doctor, and you too, my dear – very nice brand of sherry you have here, I look forward to sampling it again some time.

Me: A pleasure, Inspector – anything I can do to help.

Writing that down, I see it was a silly thing to say. Help to do what? To prove a person is alive, when I have pretty well proved that the person is dead? Well, I suppose I could have meant the other thing – *viz*: anything I can do to help *to prove beyond doubt that she is dead.*

(There *is* one way I can do that, of course. You have to smile.)

Well, they were just turning to go when -

Dew: So you think Mr Bruce Miller may have been in charge of the funeral arrangements?

Me: Im pretty sure he was.

Dew: When she lived here in London, your good wife attended Sunday mass at the Catholic church in Kentish Town?

Me: Yes, regularly—

Dew: You may not be aware that Miller was a staunch Catholic too?

Me: No, I didnt know that. (They do their homework, down at that Yard.)

Dew: Since they were both staunch Catholics, how did she come to be cremated?

9

YARD NOT SO FRIENDLY

AND IN THE silence after that, I remembered – too late – the fuss she had made when her mothers ashes arrived. Not so much grief – she didnt like her mother that much – as shock that her step-father, a non-Catholic, had gone against family wishes & had her Mom cremated.

All I could say was, I just dont know, unless – as one has heard – cemetry space in the big cities on the Coast are at a premium, & this was the only solution – all I know, Inspector, is that they are *her* ashes, in that safe.

Dew: True, one has heard about these cemeteries getting more & more congested, in my opinion cremation is so much more sensible & civilized, when you think of those bodies mouldering away—

Me: I most certainly agree.

More small talk, & they went. Polite as ever.

But I dont feel as confident as this time last night.

11 pm

Quite perturbed, & wont sleep well. Its like this.

My D. seemed fine after the Insp. left, & went down to kitchen to prepare dinner – I didnt go down with her like I usually did, as I had some tax papers to deal with.

It wasnt nighttime yet, but dark & close & stormy, & it started to thunder. I didnt bother because one time in the middle of the night it had thundered & it hadnt frightened her – then I heard, coming up the stairs – a shriek almost it was – Peter – Peter—

I raced down the stairs, thinking, shes burnt her poor hand. At the first I couldnt see her in the kitchen, the light was so dim – then I spotted her in the corner by the little window, her back flat against the wall, pale & breathing hard. I rushed to her & she rushed to me, I had my arms around her, what is it my Dearest, what is it?

Well, she didnt know.

She told me (like a child she was) that for several weeks now shed noticed she had a funny feeling as soon as she got to the bottom of the basement steps, & it got worse outside the kitchen door, then a bit better in the kitchen, but not much. That was why she'd suggested me being with her in the kitchen as much as possible.

What sort of a feeling, I said.

Peter I know its silly, she said, but its like – a sort of smell.

That gave me a bit of a jolt but only for a second, because there *is* no smell, I check every week.

She took hold of a potato she had nearly peeled, & went on with it. How do you mean, I said, a *real* smell, like drains?

Oh no, she said, its as if it was inside my head – just a bad feeling, oh I know I'm silly—

Then she makes a sharp turn towards the kitchen door, & I see her go dead white. From where we were, I could see the dark passage & the cellar door opposite, closed of course. I said, My Dearest, what is it? Then she said, in a whisper – she could hardly speak – Peter I just heard something – like a cry, not human—

I thought she was going to faint, & all I could think of was to get her back to normal, so I walked to the open door & looked out into the passage.

There *was* something there. Moving. And just then, there comes a great clap of thunder & something flashes past me & into the kitchen.

I heard my D behind me give a shriek & turned just in time to see the next-door cat – terrified by the thunder – streak up to the open window & out.

Well, I put my arms round her & start to laugh, & then she laughed as well, with relief – hystirical to start with, then calmed down. Then she looks down at her hand, and we see that in her fright she had clutched the little knife she had been peeling the potatoes with, & there was blood all over her hand.

I was upset & rushed her to the sink, she said, Oh its a small cut, she was less concerned than me. And she was right, the blood made it look worse than it was, I bandaged her hand with a clean dish-cloth & she went on getting the dinner ready.

I shut the window after the cat, in case of rain. She remarked that she had often noticed the cat prowling around in the passage, outside the cellar.

Then I sat down to shell the peas for her, then she said, Once *you* are down here with me, I'm alright. Then she kissed the top of my head, very light, a way she has.

Then she said, The Inspectors nice, but he *is* suspicious, isnt he?

Well, I said, you cant blame him, it's his job. Each conviction is a feather in their cap, and I've heard say theyre paid extra for each one, which makes some of them – I dont mean the Inspector for a minute, he's an honest man – but some of them, so I'm told, are tempted to advance themselves by manufacturing evidence.

My D – Oh no! But I must say, Peter, he's very sharp with his questions, I wouldnt like to be sitting across a table from him if I had a guilty conscience, would you?

Not arf, I said (Cockney again). If I was being plagued with a guilty conscience, as you put it, I'd have a hard time with him, trying to wriggle free with some swift lies!

Then my D looks at me, quite shocked. Peter, lies from *you*? You just paid a compliment to the Inspector by calling him an honest man – what about *you*?

Me: Me?

Her: Oh Peter, I know a dishonest man when I see one – I had enough practice with my father. I did love him – still do though I never see him, another man has taken his place – but I could never trust him the way I trust you, he'd swear he hadnt been at the pub at all, he'd been run over by a cab on his way to work, but he'd been at the pub the whole time –

whereas you— Remember the time you tried to lie to me?

Me: I lied to you?

Her: Last December when you didnt turn up at the Walpole, and you said youd had an emergency call to a casualty with a broken jaw & teeth knocked out – well, remember me saying, Peter youve gone red in the face, what *really* happened? And in the end I got it out of you – you'd been telephoned and gone to get my father out of Bow St. Police Station & didnt want me to know, thats the sort of liar you are! You are the most honest man I could ever meet.

(And she was reporting a true happening, I'd forgotten it.)

I nearly said, Oh but this is different – but I didnt. It was such a sincere moment between two loving people. She was right, I couldnt pretend *anything* with her. Except to spare her feelings.

That is why, with me & the Insp., it's different. He's alright, but its me versus him. Self-defence. Defence of me, & of her.

Her saying all that did me the world of good. And I needed it.

I had been real upset. Upset at seeing *her* upset, I mean.

And she *was* scared – thats the worst of being a timid young creature thats too imaginative. But now she's told me, & got it out of her system, she'll be alright.

And I think the cat helped.

Fri. July 8, 7.30 am

No sleep.

Im not even wondering if he's coming back. Just wondering *when*.

And Ive got to be ready. Knowing him, he'll only come back if he's got a couple of questions. Strictly routine.

Questions about something new he's got on to. What things?

Suppose it's something I wont be able to answer? Because therell *be* no answer.

If that happens, Ive got to make up my mind as to what would be my plan of action.

When in a tight corner – no, thats putting it too high, lets say – when faced with an embarrasing dilema (cant spell all that) – the truth can have its uses.

But not here. Oh no.

If only that trip to Dieppe (Dr. & Mrs. Crippen visiting Monsiur le Frenchman) had happened, & been a success – me & my D would be married by now. No use thinking on those lines.

The truth can have its uses – that keeps running through my head. Looking back, theres something on my mind that I dont enjoy dwelling on, because it could be another if only.

And here it is. The night she had that accident, was I too full of the Dieppe scheme (and of my 100% intention to carry that scheme through) – ie, should I have wiped that thought completely off the slate? Instead, should I have said to myself, This thing tonight being an accident that could happen to any husband (even a loving one) with a difficult hystirical wife, I must leave everything as it is & telephone the Police *now*?

From what Ive studied of the man Dew & the police attitude & methods, I have a feeling they would have believed me & by now—

No, that *is* an if-only, & no good tormenting myself with it, whats done is done. Whats dead cant be brought back to life.

Wait a minute. Yes it can.

MEMO Talk (assuming that Dew *may* somehow corner me).

Inspector, Ive been thinking this over, & the moment has come for me to make a clean breast of it, & Im only sorry to have wasted your time.

As you may have guessed during our talks, I wasnt always telling the truth, & in a minute I'll tell you why.

Firstly, my wife is not dead (not, at least, as far as I know) & *secondly*, she did not leave to see to a sick relative, she walked out on me for good, to join her fancy man Bruce Miller in California.

Then Dew will ask—

Dew: Why did you pretend about the relative?

Me: I didnt want to be talked of as a fool by all her friends. (True)

Now suppose then he asks—

Dew: But why did you want to make out she was dead?

Now this will be my strong card – this is where the truth, & nothing but the truth, will hit strong & hard.

Me: Inspector, I happened to know via a patient of mine, a Mr. Matheson in Golders Green, whose wife deserted him 4 years ago & never been back – just disappeared – & who wants to marry again, that you have to wait 7 yrs – yes, Inspector, 7 – before you can marry again. I know you like my fiancee Miss Le Neve, & you must have seen that I just worship her—

(Its true!)

Well Inspector, can you imagine a perfect lady like her, with the prospect of us having to wait 7 yrs before I could do my duty & make an honest woman of her? Inspector, put yourself in my place, as a happily married family man – suppose *you* had been in this situation when you were hoping to marry your present wife – *7 yrs*!

I must sit down & try to get that by heart. Not that itll happen. It wont.

That letter – from my son – keeps getting on my mind, I must have a think about it.

(Its not easy to, Ive got it so fixed in my mind that he *did* write it, which of course is a very good thing.) Thank God I went for the sheet of paper to a stationers quite a way off, without my glasses & had no conversation with the girl. I just said, I want a few plain sheets of paper & she just said, Quarto or what?

(I answered plain white – I'd never heard the word, & I remember later asking my darling what Quarto meant, & being a Sec, she knew, it's a size of paper.) That girl would never remember me. So thats alright.

MEMO – this morning take this (ie Diary) to Office –

luckily strong-box just fits into my attache-case – & in future store it in safe.

Take no chances.

Office, New Oxford St., 2 pm

Good thing I brought Diary here, I just couldnt fix my mind on business – between patients I feel a bit jumpy & this steadies me. A slack day, so I have time on my hands.

Plenty to write about too (my D out delivering prescriptions, so isnt here to notice me scribbling).

I somehow wasnt too surprised to see the shadows of the 2 of them on the glass of the door. Almost expected it. 10 am.

My D answered the door, & when she said, Oh come in Inspector – if she felt any aprehensions she certainly managed to hide it. I really think the dear girl is getting the better of her fancies.

Insp. not quite so friendly. Good morning, I have just a few questions. Not Good morning Doctor.

And that damn Serg. Mitchell with him again. Now theres a poker face. Part of his job I suppose.

They sit, my D takes her work into my private room. Insp. asks if he may smoke his pipe. Then we agree that the weathers changed, not quite so sunny as yesterday – all as per usual. But I know his tactics well enough by now to say to myself, you never know the minute.

When the Serg. put down a Gladstone bag & brought that notebook out, I made out not to notice.

But I didnt like it.

But I kept my smile on. Then the talk. And not small talk neither.

Inscrutable.

Dew: My first question—

And then he remembered something & took an old envelope from a pocket, with some scribbling on it.

Dew: Excuse me – while I remember, Mitchell, re the Copleston Case, this note should go to the Assistant Commissioner today – (then to me) would you allow the

Sergeant to type it out for me – no, not on your Office paper, just a plain sheet.

I said, Of course, & the Serg. went to my D's typewriter. While he typed – it was only three or four lines – the Insp talked about the Copleston Case, how theyre trying to prove that when she married him she did know he had a wife already & is not the innocent party she passes herself as, it was most interesting. Then when the Serg. was back with his notebook, Dew took out a folder, carefully took out my son's letter, as before. I thought, I'm getting a bit sick of that letter. The Serg. finishes his typing & goes back to his shorthand notes.

Dew: Has your son ever visited you in London?
Me: No never.
Dew: Has he ever visited this country?
Me: Never.

Then, soft & polite, another little bomb-shell. Out of the corner of my eye, I could see the Serg.'s head was up, his eyes fixed, 4 eyes, on me.

Dew: Then we've got a bit of a puzzle on our hands here. The paper on which this letter was typed, judging by the watermark and the fact that it's a quarto size, different from the American, was bought in England.

Nobody spoke for a minute. It was up to me. And something came to me – what my D had explained about Quarto paper. I said, it sounds strange but my son is a writer – journalism, short stories & such, theyre often the roving restless type – & he always uses the typewriter & favors the Quarto Sized paper which you cant get in the States & which he first borrowed from an English friend in Calif. & has run out of – so a year ago he wrote me for a packet of it, which I sent him.

Another silence.

Nobody spoke again, Mitchell writing away.

Dew: When he wrote asking for this paper, was that on the back of a Christmas card too?

He was smiling, with like a twinkle in his eye. And thats when I said to myself, Hello, he cant *prove* thats not true, & he knows it. Watch out. I was thinking this while – for

1910 JULY 8: ALBION HOUSE, NEW OXFORD STREET 133

something to do – I was fitting the cover over my D's typewriter.

Wait a minute – her typewriter. Which the Sergeant just used to type a note to the Commissioner. Commissioner my eye, it was just an excuse, so as to be able to compare the typing with the letter thats supposed to be from my son.

Well, theyll draw a blank on that. But—

MEMO. Remove old typewriter from back of box-room, walk it to Regents Park Canal & dump it. Must keep ahead of him.

(One thing occurrs to me & it interests me.) When he mentioned the Copleston Case theyre working on, & the important note to the Assistant Commissioner, at first – of course – I didnt tumble to what he & the Serg. were up to, I thought the note was a genuine message. And what did I feel, as I recall now?

Jealous. Plain jealous, because he was interested in any other case but me. I just felt like I was the only one. Not that I'm a *case*, of course. I just felt I was the only person he had his mind on.

(Crippen, youre getting big-headed.)

Then we have a bit of small-talk again. I'm getting wary of said small-talk by now, thinking, what's this leading to, and I was right.

He's just mentioning that yesterday after his pleasant glass of sherry at No. 39, he took the wrong omnibus for Waterloo Station. I put him right on that, then he weighs in.

Dew: Talking of omnibuses, yesterday evening when you left your brief-case on the omnibus, were you downstairs or on the open top?

Me: On the top, it wasnt raining after all & since I like to travel by omnibus, I sat on a front seat, to get the view.

Dew: And very pleasant it can be. By the way—

And cool as you please, he looks at the Serg. & gives a nod, quite brisk. And the Serg. opens his Gladstone bag, the Insp. leans over & fishes something out of it.

My brief-case. With the straps done up.

Me: Good gracious, thats quick work! (And I meant it) And are the 2 cables inside, & the Death Certificate?

I asked that so quick & eager, I give myself full marks. But wait.

Funny, he doesnt answer, just undoes one strap on the brief-case, then the other. And takes time doing it too. Then he holds out the brief-case, pulling it open for me to see.

Dew: When found, it was as it is now. Empty.

Me: Well, if you remember, its what we were afraid of – the thief got to some quiet place with it, found nothing of value & (like they so often do) in temper flung the contents – including cables & Death Certificate – into a dustbin—

Dew: It wasnt in a dustbin, the conductor found it on the top of the omnibus, in the front where you had been sitting.

Me: With nothing in it?

(I said it to gain time.)

Dew: With nothing in it.

(Inscrutable again.)

Me: I remember now – when I got off without the case, the top was empty except for one man right opposite me, something shifty about him. I remember wondering afterwards if *he* could have pinched it.

Dew: But he didnt, here it is—

Me: I mean that he took what was inside.

Dew: Can you really believe that?

Me: Well, I have to, since what was inside isnt there any more. (I felt what I was saying made really good sense, pride before a fall.)

Dew: Its hard to credit that a thief would pick up a brief-case, take the trouble to open it (you saw the time it took me, just now, to undo these 2 straps) find two worthless cables, a Death Certificate & some prescriptions, pocket them & leave the brief-case on the seat where it was spotted?

Me: Looks like it, doesnt it?

Dew: I'm afraid it doesnt, for another reason.

Me: Yes?

Dew: The conductor going up the omnibus stairs collided with the man you described as having sat opposite you. The man said, This has been left behind, & held out your brief-case.

Me: Open?

Dew: And empty.

Me: Of course – he'd opened up the straps, and stuffed the contents into his pocket!

Dew: Had he? (Smiling, gentle & reasonable.) Two reasons why he couldnt have. First, if he had, why would any thief in his senses go out of his way to draw attention to his theft? Second, what he *then* said to the conductor was, Get him quick, he's just got off! They were at your stop, you see, & from the step of the bus he & the conductor were just in time to see you walking round the corner. Then the conductor decided it wasnt worth pursuing & the bus went on its way. So if you were still in sight a few yards off, the man couldnt possibly have had the time to undo the straps & get at the contents & pocket them. So he cant have been quite so shifty as he'd looked to you – more the Good Samaritan, if anything!

End of speech. Still smiling. But for the 1st time, he'd got me. What could I say, *he'd got me.*

Well, thats what I said, in those words, smiling as well, & looking puzzled – Inspector, youve got me there!

Dew: Its just one of those mysteries, would you say?

Me: I would indeed, another case of truth stranger than fiction!

Dew: Thats right, the Case of the Missing Death Certificate. Well, this is a mystery we dont look like solving, eh Sergeant?

Well I thought, it looks as if I'm out of that wood. And I start in to ask him about the Copleston Case, man to man.

But the game isnt over, he's got another card to play.

Another ace. He takes something out of his pocket & hands it to me.

A cable.

For a second I thought, whats *this* twist? For the reason that the 2 missing cables from my son were by now so real in my mind that I thought, Hes found one of them, what *is* it?

He laid the cable in front of me. Here it is, pretty well word for word.

NO TRACE IN CALIFORNIA FILES OF REGISTER OF DEATH OF

PERSON YOU INQUIRED OF STOP WE CONFIRM THAT SINCE EVERY DEATH IS AUTOMATICALLY FILED HERE SHE CANNOT HAVE DIED IN CALIFORNIA STOP REGRET CANNOT ASSIST YOU BUREAU OF VITAL STATISTICS IN SACRAMENTO.

Patient just arrived, abscess, in agony poor thing.

3.30 pm

Where was I – yes, the cable.

I took quite a time studying it. Then I said to myself, How right I was to work out what to say, if I'm ever caught out.

Because here I am, caught out.

Then I thought of my D – she would have to know it all later anyway & it will have more effect if she hears it with the 2 of them present. I must fetch her.

Then something happened that I didnt like too much.

I get up, v. calm & desisive & take a step to the door. It must have looked a bit sudden, because (out of the corner of my eye) I saw the 2 of them – exactly together – they shot up from their chairs like a pair of jack-in-the-boxes, streaked past me, swung round, & barred my way to the door.

You have to take your hat off, it was the neatest bit of drill you ever saw. They were looking at me, very steady. Kind of wary, nothing more. But I knew they were taking no chances. I thought, they look as if the handcuffs will be out any minute.

Then I thought, dont be a damn fool.

Dew: Did you want something?

(No calling me Doctor.)

Me: Well yes Inspector, I want Miss Le Neve to come in, so she can hear what Ive got to say. Youll understand why in a minute.

The Inspec. looks at the Serg., & seems to loosen up.

Dew: Fair enough. Sergeant, get her, will you?

And we sit down again, the Serg. comes back & holds the door open for my D.

She didnt look too alarmed, she told me after that she

thought I wanted her to type something. I took her arm gently & said, Sit down my dear. She looked a bit aprehensive then, but I was behaving as reasuring as I could.

And when we were all sat, the interview went exactly as I had prepared it, I might have seen it in a crystal ball—

Inspector I'll make a clean breast of it, my wife is not dead etc.

The only thing was of course, I hadnt prepared for my D being present, & when I said, My wife is not dead, she gave quite a cry & sat staring me with her hand to her mouth.

I said quickly, Dont be worried, my darling, let me finish & you'll understand. Then, as I turned to the Insp. again, I thought – good thing they saw her do that, because it proves she believed in the Death Certificate, no girl could have acted *that*.

Then everything went through according to plan, with the Sergeant's shorthand scratching away.

Then, when he got to his question about, *Why did you want to make it look as if she had died in U.S.A.* – then I let fly, about me waiting 3 long years. I said, For the past 3 yrs this young person has been my only comfort.

Then, from the look on her face, with the surprise & the pain giving way to a look of love, & her lips trembling, I heard my voice break & felt my glasses steam up, it was heat from tears.

And I could tell that the 2 Scotland Yard men believed me, every word. They had to.

Not that the Insp., with his tough experience as a man of the world, didnt see the cynical side of the situation, as I would have myself.

Well, he said, a nice kettle of fish. First the lady's alive in California but no news, then she's dead & buried in California – sorry, cremated – then she's alive & kicking again in California – but still no news! He & me both laughed, though the Serg. didnt, I suppose he didnt like to. I decided not to look at my D, having the idea that she wasnt even smiling.

Then the Insp. said, Of course, I shall have to find Mrs Crippen, to clear this matter up. I said, Can you suggest anything, Inspector, would an Advertisement be any good?

He said, An excellent idea. And then I drafted one, *Will Belle Elmore communicate with H.H.C. or authorities at once. Serious trouble through your absence, 25 dollars reward.*

Then the Insp. got up, quite breezy, & goes towards the door.

Dew: Well, I'll be off on the Copleston Case—

I thought, Well, thats over, now a breather – but no, at the door he turns. One little thing, he says. I thought that mind of his works quicker than from the looks of him. Yes Inspector?

Dew: Now this letter – it puzzles me.

And from his pocket he takes a piece of paper.

Dew: Or rather a copy of a letter. In your handwriting, dated April 7th of this year, & addressed to your wife's half-sister Mrs Robert Mills, care of Mr F. Mackamotzki, in Brooklyn, New York (Mackamotzki, the name came back to me in hate, & that was a help at a sticky moment).

He reads out from the letter. Thats when I wished I hadnt insisted on my D joining us. There she sat, & without looking I knew she was as white as the paper he was looking at.

But I knew from my hands hanging loose that I look relaxed & easy. Thank God.

The letter. Since I had taken a lot of trouble putting together the first opening words, I recall every one of them.

My Dear Louise & Robert, I hardly know how to write to you of my dreadful loss, the shock to me has been so dreadful that I am hardly able to control myself. My poor Cora is gone.

And so on. The Insp.'s voice was as flat as a schoolboy reading a reciting piece, & somehow that made it sound – well, worse, I wasnt too comfortable listening, especially when he got to the sentence, *Never more to see my Cora alive nor hear her voice again, her remains are being sent back to me, and I shall soon have what is left of her here.*

Then the Insp. made a remark, again so flat it didnt sound sarcastic at all, holding out the paper.

Dew: This copy isnt exactly like the original at the Yard in that it hasnt got the black edge all round it.

It was then I heard a sound, a sort of quick breath & a gulp. It was my D, & I still darent look. She was suffering & I would have liked to do the suffering for her. But no chance, I could only try & keep my end up.

Dew: How close was Mrs. Crippen to this half-sister?

Me: Oh, very close.

Then the Insp. gives me a sharp look, & the Serg. looks up from his scribbling & looks at the Insp. What have I said, I thought, but for something to say, I went on.

Me: Very close, thats why I felt it my duty to write & break it to her.

Dew (sharp again): Break it to her that her half-sister had passed away?

Me: Thats right—

I hear my voice trail off. And I say to myself, Crippen youre getting into a muddle.

Dew: But youve just told me she hadnt passed away, that at the moment of speaking she's alive & in the States?

Me: That is correct.

Dew: And you just stated that your wife was – I mean *is* – that your wife is very close to her half-sister in Brooklyn. When you wrote this letter to the said half-sister, didnt it cross your mind there was a very real danger there?

Me: Danger in what, Inspector?

Dew: In informing her that your wife had died. And *not* died in London.

I recall that I didnt fancy the sound of that.

Me: Thats right. I informed her my wife had died in California. What was the danger, Inspector, in that?

Dew: In you *pretending* your wife had died.

Me: Thats right. And Ive told you why I pretended – so I could be free to do the right thing by this young lady here.

Dew: But with you knowing your wife was alive in America, & that she was very close to her half-sister – your own words – didnt it occur to you that after the half-sister getting the letter from you, any minute she might get one from your wife, which would – to say the least – have puzzled her?

I thought, get moving Crippen, work fast.

Me: Inspector, when I said they were very close, I was speaking of blood relations, only. My wife – to put it mildly – was a quarrelsome person – & must be, still – & the fact is that before she joined me for good in London, she had – for all sorts of reasons – burnt her boats with most of her relations, she just wasnt on speaking terms. There was no likelihood of contact.

Dew: And yet you wrote to the half-sister about the death?

Me: I felt it was a sort of family duty.

Dew: I see. By the way, Doctor, whose ashes *are* they?

Another jolt. Inspector, I said, I have to be honest, theyre Mrs Mackamotzkis.

Then the Insp made a remark which I consider humorous, though not quite tasteful under the circs. I see, he said, mother deputizes for daughter.

Then he turns again, very brisk.

Dew: Well, back to the Copleston Case – while I'm gone do you mind if Sergeant Mitchell waits in the Office across the corridor, I may want to phone him about that, if you dont mind – thank you very much.

And he went, then my D showed the Serg. into Dr. Masters office opposite. Luckily his Sec. was at the house seeing to the funeral on Sat., she would have thought it funny to see the Serg. sitting opposite her in her office with the door open so he could keep an eye on my door.

In case I left.

But thats only natural.

Now I recall something. When the Inspect. said, Thank you very much, & went, there was a look in his eyes that wasnt there before. What was it?

It seems to me Ive seen it somewhere else, though never on people talking to *me*, or looking at me.

I remember – I saw it once when caught in a crowd at Liverpool St. Station & the Stationmaster etc were greeting the King, & Queen Alexandra, on arriving from Sandringham. He & the group were looking at the K and Q with *respect*.

That was it, *respect*. Mind you, people seem to have quite

liked me, but in a sort of tollerant way, I could hear them saying behind my back, little Crippens a nice chap even if he doesnt add up to much. (I have an idea that folks have found me easy to dismiss from the mind.) But the Inspector looked at me with respect.

Once my D & I were by ourselves, I turned to her.

Now when I had come out with all that about loving her, she had looked soft & happy-sad, sort of – she had to. But now, *she* had a different look too. She looked *scared*.

Yes, I know I'd seen her scared before – & out of her wits, night before last, from that cat & the thunder – but this time she looked scared of *me*.

I put my arms round her & said I was glad she'd heard all that about me & her, sat her down & even gave her a sip of the emergency brandy like I had when her father was so sick, & I said, What is it my Dearest?

My D – Peter I dont understand – whats happened – why did you tell him shes still alive when you have a letter from your son – you *showed* it to me – telling all about it & that it enclosed the Death Certificate that you left on the omnibus – what is it all about?

Then, telling her now the details of all that – typing the letter myself, copying my sons signature, leaving the empty brief-case on the omnibus on purpose – I realized that when I had made a clean breast of it, in front of her, to the Inspector (about there being no death) she had been completely at sea because she could not *believe* I had not been telling the truth all along.

My D: Peter, I couldnt believe my ears – you had been telling *lies*.

Me: But I *had* to tell him lies, dont you see, so we could get married sooner—

My D: But you had lied to *me*—

Me: But what would be the point of telling you any different? If I'd let you know from the start that I was misleading him & that she *is* alive, from what I know of you it would have worried you to death – youd have been awake all night, afraid he might find out there was no Death Certificate etcetra – you do see that my Dearest dont you?

And once again that was dead easy for me, to say all that, because it was the truth – 100% what I felt.

But coming out with all that to her, about making up the Death Cert. – what were *my* feelings?

Well, quite different from with the Insp. – with him it had been, all along, like a kind of game, a fight almost. But with her I was like a schoolboy whos been caught & has to own up. Except for her scared look, I'd have thought – almost – that I was in the witness-box, like a criminal.

Eg, when – I remember – when I told her how I had copied my son's signature, she interrupted, quite sharp for her. Forged, she said, you mean *forged*.

Well I had to say, Yes if you look at it like that – but she didnt seem to understand, just looked as if she couldnt believe her ears. She kept remembering things – but the obituary notice you put in the *Era* – the memorial cards with the black edge you sent out—

Sounds ridiculous, but it was almost as if she was *accusing* me.

I said at the end, My Dearest, you must remember that I did it for *you* – for *us*, for our future! I know, she said, I know— (She just had to believe that.)

Then – oh yes, I made a mistake.

Me: Did you notice the look he gave me just now when he left?

My D: Yes I did.

Me: It was a look of respect.

My D: Respect?

Me: For me, for having planned all that & within an ace of getting away with it.

(Shouldnt have said that, though of course it was what I believe – what I should have said was, respect for me for my honourable intentions towards you, my darling.) As it was, she just looked at me.

My D: Getting away with what?

For a second – it was like – yes, I know, like being on thin ice & hearing a crack. Get off the ice.

Me: Well, getting away with convincing him she's dead, & getting that bit nearer to you-and-me being married.

My D: There may have been respect, Peter, but I saw something else as well. I did—
Me: What?
My D: Suspicion.
Well, I stared at her, & I gave quite a chuckle, it was such a funny word to come out with. Suspicion, I said, why my Dearest, you sound like one of those Sherlock Holmes things – suspicion of what?
My D: I dont know – I just felt – (I dont think she knows how she feels, poor pet). You dont think – they might report you – or whatever they do – for fraud – no, I dont mean fraud – sort of misrepresentation of facts, like perjury—
Me: Perjury, my dear, is when youre in a court giving false evidence under oath, now just calm yourself & get on with those prescriptions, theyre badly needed & we dont want patients dying on our hands.

I was trying to jolly her along like a dutch uncle, & I think I managed it pretty well. I must continue the good work.

She is back from her errands & typing at the minute, quite herself. She just couldnt bear for me to get into any sort of trouble – the same as I think of her, & hate to see her fretting over the slightest thing. I still sometimes take a sniff at the eau-de-cologne, she has teased me about it. We love each other.

We're going out for a sandwich. Then back to work, her to her typing, & me to – well, waiting for patients, as usual.

5 pm

No, not as usual. She keeps staring at her typewriter & biting her lip, then she would feel my eyes on her & force herself back to work. For the first time ever, I felt there was a sort of wall between us.

And if it stayed like that, it would break my heart.

After 10 minutes, to cheer her up, I repeated to her what the Insp. had said about her at lunch that first day. I told her his exact words – What he said about you my Dearest was

She is such a charming young lady she deserves to be happy, if I had a daughter Id want her to be like Miss Le Neve.

Well, normally she would have blushed from the pleasure of such a compliment but she just looked at me, sort of hesitating & said, Did he? & went back to her typing.

And that funny look meant she was thinking, Is it true, *did he hear the Inspector say that, or is it all made up like the other things?*

I who have never told a lie unless absolutely necessary. I cant bear to think she doesnt trust me.

A minute ago, ran into Dr. Rylance on my way to W.C. He said, Crippen you look worried, I said, its these policemen, the latest is they keep on pestering me about that Drouet Institute trouble years ago, as if I was to do with it. I thought to myself, that sounds all right.

I can hear the Insp. outside calling, I'm back Mitchell—

5.30 pm

Steady steady. A good think.

When the Insp. came back into Office, the Serg. behind him (Serg. just come from Masters office), I noticed a difference in the Inspec.'s appearance, I couldnt tell what it was, except he was somehow not so spick & span as before – grubby almost.

He asks my D if she minds leaving us alone, she goes, then he asks if I mind him washing his hands at the wash-basin. I said, Make yourself at home. He seemed to have been doing manual labor or some such.

I ask them to sit down, quite calm, but I knew something was coming & braced up for it. I said – in my mind – over & over again – I am an innocent man, *I am an innocent man*. Just as well I did.

Dew: I'm afraid you wont be very pleased with what I'm going to tell you.

Me: Oh? I'm sorry to hear that, Inspector—

Dew: Ive spent the afternoon up at your house.

1910 JULY 8: ALBION HOUSE, NEW OXFORD STREET 145

(I am an innocent man, *innocent*.)
Me: Youre right, Im not at all pleased. Isnt that against the law?
Dew: Im afraid not, I have a search warrant.
Me: May I ask what you were searching for?
Dew: Anything that might help us to trace your wife.
Me (*And this took some asking*): And did you find anything?
Dew: About that old typewriter in the attic – is that the one you wrote your sons letter on?
I had pushed that typewriter into a dark corner behind a lot of stuff. He's thorough. Is he trying to break me down? I'm tougher than I look.
Me: Thats right, I thought it would be too risky to use this one.
Dew: Quite right.
Then he takes a handkerchief from his pocket, lays it on the desk, takes tweezers from pocket, unfolds handkerchief with tweezers.
Inside handkerchief is something with dried blood on it.
I thought – cant be, I was so thorough. Then I remembered – it was the dish-cloth I had tied round my Ds hand last night when she cut it peeling the potatoes.
I explained that to the Insp. – the cuts still not quite healed, I said, she can show it you, in the other room – I remember throwing that into the rubbish-bin under the sink. Where did you find it?
Serg.: Under the sink. Half behind the bin.
Dew (*quite genial*): You must have missed your aim.
Me: Must have.
Old typewriter, dish-cloth, he's been through the house with a microscope.
And found *nothing else*.
Thank my stars I moved the Diary out. Just in time.
Then what happened? Let me see – I asked them how business was – we all three had quite a chuckle about that – then the Insp. said they were catching a train to Leeds, for a big forgery trial the Yards been involved with, a long tedious job but it looks as if the verdict tomorrow (bound to be

tomorrow, he said) will *be in our favour*. (I was interested in his way of putting it).

Now I come to whats on my mind. We were chat-chatting about No. 39. (Now I come to the important part) That it was a big house to run and how clean it was kept (they should know). Then I said what had been in my mind for several days:

Me: You see, Inspector, with all this coming out about me being misguided & pretending that my wife was dead, & all the discussion about that, Miss Le Neve has been feeling more & more uncomfortable in the house—

Dew: Thats very understandable.

Me: And I'm seriously thinking of buying a flat. (It *had* been an idea, for a minute.)

Now I'm coming to the part thats on my mind.

We chit-chatted, then he talked about him & Mrs. Dew having bought their own house a couple of years back, then we started talking about which house-agents are the best – I was only half paying attention, my mind being on my D next door & how I was to keep from her the house having been searched again – then for something to say I asked – let me get this straight:

Me: When you buy property through a house-agent, does the buyer pay the commission to the agent?

Then the Insp. says something about the commission being split, or something – then he says (looking straight at me) he says, *In the cellar*.

I stared at him. The cellar? I said.

He smiled then, as if he was puzzled, & he said Yes, I was answering your question, I have an idea the commission is usually split between the *buyer and the seller*.

(He had swallowed the word *and* (the way people do) & so it had sounded like *buyer in the cellar*.)

Me: Oh yes of course—

What else could I say? I could have kicked myself, if I hadnt brought up the subject of estate agents, it wouldnt have happened.

I was so knocked off my perch I cant rightly remember what happened then.

1910   JULY 8: ALBION HOUSE-HILLDROP CRESCENT   147

Yes – he & the Serg. looked at each other.
Theyll be coming back to the house. Not now, nor tomorrow, they wont be back from Leeds till tomorrow eve.
But theyll be in there the next day, Sun. the 10th.
Got to get away.

39 H. Cresc., 9 pm

Brought strong-box home, Diary in it. (NB Must burn Diary.)

Sent my Dearest to Mrs. Masters, for condolences before funeral in the morning, so she wont notice me humped over blank page. (Which I have been since I slipped down to have a look at cellar.) The old chandelier has moved again, coal disturbed, but must have been v. casual, brick not tampered with.

Oh dear, I dont like what happened just before she left for the Masters house.

She was sitting at sitting-room desk sorting out household bills, when I heard a sharp tearing noise, looked round & she was tearing something into small pieces, then she threw them into the wastebasket almost as if she was angry.

While she was upstairs getting my bedroom slippers, I got up & looked in the basket at what she had torn up. Her eyes had caught two or three cards left over in a cubby-hole of the desk. I saw a strip of black. Memorial cards.

I did all that for the best. She's got to see that.

Steady, steady. Think.

Theres a lot to work out, one thing at a time.

Out of the country, & her with me.

Dieppe. (I know Dieppe, Dieppe doesnt know me, suits me!) Night boat tomorrow night July 9, 11 pm.

First, my D will be back from Mrs. Masters any minute. MEMO of talk with her, & a difficult one too.

My Dearest, I dont know how to break it to you, but believe me Ive thought this out for well over an hour since the Police left – yes my Pet I know it sounds alarming, calling them the Police, more than calling them the Yard –

but Police they are, make no mistake about it & this afternoon they showed themselves in their true colors.

My Dearest, theyve had the impudence to go up to No. 39, behind our backs & ransack it from top to bottom, I said how dare you, what were you looking for?

Papers, they said, anything which would help us to trace her whereabouts in California. And of course they found it, at the back of my drawer. I'd forgotten it – you remember when I told you she'd written a note asking for the rest of her things. Well theyve got her adress from that & they hinted theyll be cabling her to pay her fare here. To give evidence. Against my D.

My D, dont carry on so. I'm here to protect you – the fact is that when you were afraid I might have done something unlawful in stating to two Police Officers that she was dead, & I said Nonsense – *well I was wrong.*

It *is* a misdemenour & I can be prosecuted, then it would all come out in the papers about you living here, & I cannot watch your good name having mud thrown at it – now listen carefully.

For you & me, the only hope of happiness is to get out of the country, tomorrow night. And then sit & work out how we can travel somewhere & start life 100% fresh, you & me, & get married—

Never mind about bigamy, nobody will ever know & in a case like this *nobody* with any human feelings would ever look on it as even a misdemenour.

(And I feel thats true, *with all my heart*) & now, my D, shall we think about packing?

(NB Pack haircutting scissors.)

11 pm

Well, the female sex is extraordinary.

I expected hystirics at the least. But to my amazement, & relief, she took it all a 100 times better than I expected. Sat almost calm, her hands together in her lap, didnt speak until I'd finished.

Then she said, Ive been expecting something like this for days almost without knowing I was – just a vague pressentiment – and now its on us & we can do something – I feel better. Peter, I have a feeling we shall be alright, because I love & trust you with all my heart.

I nearly cried when she said that. She will make me a wonderful wife.

Get my D to sew 4 rings, 2 brooches into lining of one of my undervests.

About tomorrow. My D will be home, going through everything & packing & waiting for funeral.

Get to office at 9 am. First, make request to William Long my dental mechanic to go out straight away & shop for articles for boy aged 15: 2 shirts (guess at size & tell him) 2 collars, brown suit, buttoned boots, flat cap.

(He will not think it so strange, a year back I asked him to shop for some down-&-out youths at hostel run by a clergyman patient.)

Then a note to be left for him. *Dear Mr Long, Will you do me the very great favor of winding up my household affairs. There is £12.10 due to my landlord for past quarters rent. I cannot manage about Valentine the French girl, she should have enough saved to get her back to France. Kindly leave keys with landlord etc. Thanking you with best wishes for your future success & happiness. Yrs faithfully—*

Then note to be left for Dr. Rylance. Good thing I told him that Scotland Yard has called on me to find out if my wife had any estates to pay taxes on. I can use that. *Dear Dr. R – I find that in order to escape trouble I shall be obliged to absent myself for a time.*

Then explain to him that I have already paid Goddard & Smith £10.12.0 rent, in advance up to Sept 25, also that there will be several paid bills to enter into file on my desk, & that he will find key of desk in upper drawer of little cabinet in Coulthards office.

At 10 am to Bank, withdrawals. 1 oclock, Masters funeral.

Midnight

Now this *is* a quandary.

We had done quite a bit of packing & I sent my D to bed (pretty tired, poor pet) & told her I'd light a fire in the study to get rid of some old papers. (Gave her something to make her sleep, thats the good thing about being a doctor.)

And here I am, sitting with the fire lit, & staring at it, & then at the two thick books in front of me.

This Diary's come to be a part of me. And I cant bear to destroy it.

And yet nobody – *nobody* must get hold of it.

And yet – imagine in a hundred years time—

I must think. Wait a minute. Yes.

That dinner party, when the Masters came to us – some weeks back – when Mrs. M. hinted that the Dr. had knowledge of people in high places that could be of great interest in years to come.

Thats it. Get a plain piece of paper (English watermark, which is OK, nothing that could catch me out!) and pot of paste. But first – I'll examine my pigeon-hole of past letters.

A minute later Found exactly what I need – a letter from him to me, signed S. T. Masters. Write on the plain piece of paper, in capitals – IN DEFERENCE TO ROYAL FAMILY, NOT TO BE OPENED TILL – Till when? Nice round number, say 75. NOT TO BE OPENED TILL 75 YEARS AFTER MY DECEASE.

Then cut out lower part of Masters letter (the part with his signature on it), paste it under that inscription.

I recall well that his desk is in a sort of study, like mine, bathroom just outside. After the funeral, with the sitting-room full of the mourners, I will pick up my attache case in hall, nip up as if to bathroom, get to desk, deposit strong-box at back of a bottom drawer.

Knowing Dr Masters wife & family, that wish of his will be respected.

Let me see, were at 1910. Add 75 yrs.

1985. Looks funny on paper. I dont like to think of my D as old, but she wld be 102. Me, 123, & much too far gone to bother about.

But I hope therell be a certain amount of interest.
Fire is nearly out. Glad I didnt feed it. I'm tired. (Put out black tie for morning – ditto for *mourning* – both!)

Sat. July 9

My D slept, of course – so did I, in snatches, but too excited to go right off. Filled with love & admiration for her, so sensible & prepared for hardships. She's dressing for the Funeral.

2 pm

Funeral all went according to plan, Diary safe at bottom of Doctor's drawer, safe till 1985!

Oh, but I cant stand to part from this writing things down, the releif of it! I know what I'll do – I'll keep it up, only as if I was starting a Diary *from scratch*. Starting tomorrow night. If this next part, following from here, survives till 1985, they can tack it on to the first part above. For neatness.

But from now on, I must be careful not to put anything down which might be awkward in case of my privacy being invaded. In view of that, must be careful to establish my innocence by making statements in regard to said innocence – will put those statements IN CAPITAL LETTERS.

A new life!

10
THE NEW LIFE

Hotel de la Gare, Dieppe,
Sun. July 10 1910, 10 pm

EXCELLENT CROSSING CALM moonlight night. Lucky for me, I had work to do on the boat. (Thanks to the UNWARRANTED PERSECUTION OF ME BY SCOTLAND YARD CONCERNING MY MISSING WIFE.)

Unusual work. I had to sit my Dearest down in our cabin, in the middle of the night crossing, unpin her lovely brown hair, & cut it to the shape Ive been used to cutting William Longs hair, & the others. (I did that in another existence, because already Albion House is far away, & No. 39 does not exist.)

Sad to see those gleaming tresses on the floor, & then to sweep them into a newspaper & pitch them through the port-hole – but it was nice (somehow) to be bending over her with my waistcoat brushing her shoulder, & to see her in the little mirror, turning from young lady into little boy. A loving feeling, quite pleasant.

And she makes a nice little boy, about 15, William Long had made a good job of his shopping – the brown suit fitted well on her hips – rather tight – & the boots just right, a little big. She had been very nervous of the idea of course, but it was funny – once it was in the process she got interested & amused, as if dressing up for a fancy-dress do. Looking in the glass after she'd put on the Eton collar, she said, I look like *Tom Brown's Schooldays*!

Then I shaved off my moustache, shortened my straggly hair, & though I couldnt see too well what I looked like without my glasses, she said it made a lot of difference. I made her practice walking up & down with longer strides, & crossing her legs after sitting down. Quite a pantomime.

Mr. John & Master Robinson. Father & son. I think of the day she said to me, Youre my father now, & I feel quite a lump in my throat. Little would I have thought that this would come about.

Standing in the crowd waiting to disembark at Dieppe, 8 am, I whispered to her that in mid-Channel I had lost a Future Wife & gained a Son, made her smile. I could tell she was thinking, anythings better than that waiting – that uncertainty.

Funny, she is such a slim graceful girl in or out of female apparel, & yet when I walk behind her now I can see that she is female – I mean Ive never thought of her being broad in the beam, but in those tight trousers she has got hips alright. Looks nice to me but they mustnt get any bigger, not on this trip!

Mon. July 11

Waking up this am, I got quite a shock, for a moment I thought I was in the wrong room. I mean, with turning & seeing this head on the pillow next to me, cropped so close!

Its awkward without my glasses – long sight not too bad, but cant read except in bedroom, but outside I can always get my D to read things to me if necess.

Hotel des Ardennes, Brussels
Thurs. July 14

Registered as *Robinson Merchant & Robinson, Fils.*

Have made inquiries here, earliest Atl. sailing is for Canada, in 6 days time, Antwerp to Quebec, S.S. *Montrose*.

Suits us, I dont fancy the States much any more, with memories I dont care for, plus the fact WE MIGHT RUN INTO A

CERTAIN FEMALE PERSON CALLING HERSELF C. C., stranger coincidences have happened & I would hate my D to be subjected to that.

Ive booked a cabin, & we must make these next few days a bit of a holiday. Theres the big Brussels Exhibition. We can sit in the sun, Dad can have his glass of red wine & Master Robinson will have his lemonade – no, this is the Continent, they think nothing of schoolboys drinking wine, *he* can have a glass too.

From day to day, I can almost believe it is my son, I must have an imagination. I keep thinking of Canada (good mixture, to me, of best of U.S.A. & Gt. Britain) where with luck I may *have* a son, & my D too.

A new life.

Sat. July 16

Nothing to report after two days in Brussels.

Exhibition v. interesting, everything on show, from railway-engines to newest ideas of cooking with gas stoves. My D (after she had studied the latest typewriters, a busman's holiday for her) was v. interested in the stoves, with a view to our future domestic life in Canada.

Each to his subject – I went straight to a booth marked 'The Dentist of the Future', and of course she was with me, watching this Belgian dentist (not faked) extracting a tooth in some new way from a poor little girl, paid to sit there but in some pain. He had a modern method but had no concern at all for his patient. My D whispers to me, That child needs you, with that magic touch!

Her remark gave me pleasure – I am good with pain, & will be again, professionally in the new overseas life.

Had to scold Master Robinson a couple of times for sitting with fingers locked together under chin – Ive coached him to sit with each hand holding a knee. He is v. willing to learn.

Tues. July 19

Nothing to report indeed!

Strolling round on my own this morning (my D doing a bit of washing in the room) I happened to pass a big newsagents & glanced at the stand. I stood & stared, couldnt believe my eyes.

Looking me straight in the face was an enormous photo of my D.

My favorite photo of her, but enlarged to cover the *whole front page of the (London) Daily Mirror.*

I bought it, folded it over so the front page was hidden, & walked back to the Hotel in a trance. In the room (thank God) I find a scribble from her that she has gone for a stroll too, to get the benefit of fresh air & sun. I sit down, reach for my glasses, & settle down to a read. A special sort of read.

Daily Mirror, Fri. July 15. (4 days ago, & 6 days after we got away.) Under her dear face – *Miss ETHEL LE NEVE, missing in connection with Camden Town Mystery, may be describing herself as Mrs. Crippen, wanted by Scotland Yard.*

Inside – *BODY IN CELLAR MYSTERY. Since on the 13th Inspector Dew discovered the remains believed to be Mrs Crippen, the house has been closely guarded – last night it was in complete darkness except for a light in the basement,* etc etc.

This terrible thing is in front of me, & out of window I can see the sun on the houses, I cant believe it. Everything goes swirly.

Then I turn over, & heres a full page again – her (Belle E.) underneath, and the top half a huge picture of *myself,* staring at me.

My first feeling is as if all my clothes have been blown off me & I am standing on a platform in front of a crowd, in the nude.

To somebody whos never seen their name in print – except in a list at a charity dinner or when I got my Diploma – to see yourself in a *headline* gives you a very strange feeling. Like being drunk.

Its as if you are reading about somebody else & then it hits you – its *you*! And the thought of all those people taking the trouble over you – the newspaper reporters, the printers up all night, the newsboys – all to tell the world about you – Heavens above!

Then I have one of those quick visions – like a lightning picture on one of those nickelodeons – that come not only to saints but even to ordinary folk like me – a picture of me standing in a box at Covent Garden & the whole audience on their feet & applauding. But it isnt Caruso, its me. And though they cant be applauding, its attention. Not just from a whole theater, but from the whole world. Its shocking, but at the same time makes you feel – as I say – quite squiffy.

Then my second thought is – what a terrible enlargement of my Photo, newspapers must do something to them to make them look worse, those pebble glasses glaring at me, enough to frighten a grown person—

And my name underneath, in thick print. *Crippen, the Wanted Man.*

MY GOD, THIS IS UNBELIEVABLE, A BODY IN NO. 39 BUT THEY CANT THINK ITS TO DO WITH ME, THEY *MUST* TRY AND TRACE THE FORMER TENANT, THIS IS TERRIBLE – HOW CAN I PROVE IM INNOCENT?

Once the shock has worn off a bit, I study the small print. My goodness, the things they can put together in the paper, you only realize it when your own turn has come.

Mrs. Crippen's Stage Career (Thats a good one! with 4 Photos in costume.) *Daughter of Polish Nobleman.* (Thats better still) *The woman Le Neve has been seen in Edinburgh.* (Thats good, my D in Scotland, though I dont like her being called the woman.) *Crippen described as being in various seaside resorts, Brighton, Bournemouth, Ramsgate.* (Thats good too.) '*I heard two terrible screams from No. 39 during Easter*'. (But we were in Dieppe, with house empty!)

Then – *Mr. Crippen was rather a short gentleman, somewhat stout and bald, who said he had felt his wife's death acutely.* Now that description is just rude. (Not even *Dr.* Crippen.)

Then words like *ghastly, grisly, horrifying details, fiendis*h

– I'll now tear the ghastly grisly *Daily Mirror* up into small peices & drop it in the waste-basket.

Thank heaven my D wasnt with me when I saw it. Fact remains that it would be v. bad if she overheard English folk gossiping, so I best break it to her gently, over a glass of wine, that a STRANGE BODY has been found in No. 39 – & now I think of it, didnt she & me (*she* being B.E.), when we moved into the house, notice a FUNNY SMELL in the basement? So lets hope theyll clear up that mystery.

We're off to Antwerp 1st thing in the morning, & not a minute too soon.

S.S. Montrose, Wed. July 20

Steaming out of Antwerp, what a blessed relief! My D & I stand watching the ship creep slowly away from the Quay.

Goodbye to the Old World, to the old life! My first sea trip since N.Y. 8 yrs ago, & my D's 1st ever – can't count Dieppe – I will see she enjoys it.

A nice boat, not too grand, not too small & fairly empty for this time of year – nice cabin. Master Robinson behaving well, helped by slight roll of ship.

Thurs. July 21

Got *4 Just Men* (E. Wallace) out of Ships Library, far-fetched as usual but passes a couple of floating hours v. agreably. Its about a murderer whos got a warrant out for him, & £1000 reward. Now thats a lot of money.

Yesterday evening, an awkward moment. But amusing.

My D walks out to go to the toilet, opens a door marked Gents, walks in & there are 3 men with backs to her, standing at the urinal. She gave one look & scuttled out, & off to a proper W.C.

Once I made sure nobody had noticed anything peculiar, I had to laugh, the poor dear confessed she had no idea that men *stood* on such occasions, & she'd even been wondering what the men in the Dieppe streets were doing behind those metal circular partitions with their feet showing.

She was quite the blushing schoolboy, I told her that one of these days she'd be the blushing bride. We had a quiet laugh over it.

Fri. July 22, 10 pm

My D asleep in the Upper Berth.

Beautiful weather, Herring Pond living up to its name, you almost feel youre on a cruise. Moon last night, my D & I walked round & round, then sat in a secluded part behind the lifeboats, where I could hold her hand & we could watch the beautiful silver lights the moon put on the water.

We're not bothered by people, as everybody seems to keep to themselves on a trip like this. That suits us. We got talking to one Canadian, I said I was a widower & that my boy & I were going to settle in Canada, the man said it was a great wide open country full of opportunities.

And I believe that. I intend to work hard as a Doctor, & I still have the knack of being liked by patients. We are going to be alright.

Sat. July 23, 10 pm

Something happened this afternoon that gave me a real shake.

The 2 of us were on one of our little strolls in the sun, & got to the very quiet back of the ship as usual. I was pointing out something in the water some distance off, & wondering if it was another boat, & in pointing it out I put my arm around my D's shoulders (no more than that, I am naturally v. careful in public) & then she (as has often done in the past) takes my hand & gives it a little squeeze.

Well, suddenly I hear behind us (the wind carrying the sound) a man say (quite distinct), I wonder if that boy is his

son? And he gives a kind of chuckle & moves on.
It was a minute before it dawned on me what he was getting at, thank God my D didnt latch on, she just said, What did he say? & I said, He said something about the *sun*, & we walked on.
But I felt sick to my stomach. The idea of a stranger taking me for one of those filthy pederasts you read about in the Sunday papers!
I'm still v. angry.
And more so, through having noticed that strolling with this fellow was the Captain himself. I would not like to have him thinking there could be such a person on his Passenger List.

Sun. July 24, 10 pm

My D already asleep, & I can hardly keep my eyes open, its this splendid sea air.
A very nice thing has happened.
The Captain (Capt. Kendall) cannot have taken any notice of that fellow's remark, because lo & behold, a message from him to say he is asking a few passengers every day for sherry etc at 6, & could Mr & Master Robinson call on him tomorrow.
Now Ive always understood that this is quite an honour aboard ship, & I feel gratified. My D is nervous of course, but by now she has so trained herself in her *role* (as they say in the Theater) that she will pass muster, & anyway (as practically a schoolboy) she will not be called upon to converse.
Every happy day, we stand at the back of the ship & watch the old sea move away from us & the new sea approach, all the old things moving further away back into the sky (my long sight as good as ever). Then we stroll to the front (my dutiful son holding my arm) and the New Life nearer & nearer.

Tues. July 26, 1910, 6pm

The little party went off splendidly, only three other guests & the Capt. a v. genial chap, handsome, clear-cut, telling good jokes.

Sherry for me, lemonade for the Youngster. The Capt seems to take a real interest in me (not just polite) & asking how long we will be in Canada etc. I told him I was a Doctor (who had been stuck in the North of England) looking for opportunities. He was another who told me I'd be sure to find them in Canada.

A very interesting thing happened – a Junior Officer came in with a typewritten message, one of the others asked what it was & the Captain showed it to us – quite proud he was. A message to say would he contact a firm in Quebec re supplies to be shipped back to Antwerp next trip.

Somebody said, in a joke, how had he got the message – had a fish jumped aboard with the message in his mouth, yet the paper wasnt wet! The Capt explained its this new wonder they call Wireless Telegraphy, meaning they can send a message to a ship with no telegraph poles.

It is what they call a Marconi Wireless Short Distance Installation, & I was introduced to Officer Llewelyn Jones, who is called the Marconi Operator.

Just through what they call the Air Waves. Beats me. Uncanny. But what a boon to Civilization!

The Capt. has asked us in again today at 7.

Dipped into *Pickwick Papers*, always liked that, wholesome fun & educational into the bargain, & now must smarten up for the sherry (& merry) hour with the Capt. He likes us.

A v. nice trip, I dont like to think of it ending.

Sat. July 30

Another beautiful day, for our last of the trip, & my goodness how I needed it, after last night.

Went to bed after I read a bit, my D was already in the Land of Nod – when I think how nervous she had been, in that Old Life – & then I got into my bunk.

I must have gone off to sleep, then woke up with a start. I wasnt even dreaming, so no question of a nightmare, just woke up, in the dark.

And I knew that in this small cabin, in the dark, there was somebody standing. I couldnt see him, & yet I *could* – his moustache, watch-chain, pipe, thumb in waistcoat pocket, smiling at me. Inspector Dew.

HOW CAN I CONVINCE HIM THAT I AM INNOCENT?

Bit by bit my eyes got used to the faint light of moon, & then I could see there was nothing there. And yet he *was* still there. (How can I explain it? Uncanny but no more uncanny than that Wireless Telegraphy business.)

Then I got up, dressed quietly & found myself in the writing-room. Most lights were out, it was like a ghost-ship. And I sat down, must have looked like a ghost myself.

I feel fine now, so how can I believe I felt like I did, then?

It was as if all the confidence I have had up to now – the faith in myself & in my Dearest & our future away from the horrors of the *Persecution* – as if all that was a fake & I had been fooling myself, & this (which I was feeling) was the real thing.

The real thing being – that everything was up with me. I had nothing to go on, of course, but I even said it aloud to myself, *Everything is up with you.*

And all I could think of was my Dearest, the shame for her. And I sat there I felt, Im going to jump overboard, this is the time – they talked about *as easy as falling off a log*, this is easier.

After that I must have dozed off, & I went back to bed like a sleepwalker thats woken up. And this morning I was fine.

Did it happen? *Yes*, because I had gone to a desk under the one light & wrote a note, its in front of me now, I must have had the sense to crush it into my pocket before going to bed.

I cannot stand the horror any longer, & as I see nothing

bright ahead, I have made up my mind to jump overboard. I know I have spoilt your life, but I hope some day you can learn to forgive me—

Could I have written that? Well, I did. Must forget it.

Sun. July 31

Writing this at 8 am on deck, on my knee, in the sun. Ship at standstill at Father Point, off Quebec. My D only just up, a little tired after being kept awake by hooting of fog-horn during night (beautifully clear now).

Quebec only 12 hours off. Have just been lent binoculars to see glimpse of Promised Land across St Lawrence River, land just near enough to see big crowd at quay-side, the Canadians must have very little to do to be having a look at a ship this time of the morning.

Everything packed (not much to pack). The Capt. calls out that we dock this evening, weve halted here for mail & formalities. He even stopped for a chat with me, I must say he has been v. attentive.

I like these formalities aboard ship, hes standing at attention next the gangway, waiting for some official.

My D looks rested & with roses in her cheeks, better than for a long while. I ordered a glass of sherry, & she's taking a sip from my glass for us to drink to the future. And heres to it – to the Future.

12 noon

Can it be the same day?

Well, the Official arrived up the rope ladder (peaked cap & papers). I heard somebody say he was the Pilot. He shakes hands with the Capt., looks round to see everythings OK with the ship, & as they go off to the Capt.'s cabin to see to business, my Master Robinson comes round from behind some life-boats & nearly barges into the two of them. The

1910 JULY 31: S.S. MONTROSE   163

Pilot gives Master R. a look, lifts his cap & goes on.
An Officer comes up to mention to me that the Capt. would like a word. I follow him. In the Capt.'s cabin, the Capt. & the Pilot.
Then, just as I was thinking – I know that face, even without my glasses – the Pilot says, Good Morning Doctor.
Inspector Dew.
All I could think of was, that in London that last day, (& me waiting for him to) he never called me Doctor, not once. And here he is, calling me Doctor.
I answered him with, Good morning Mr. Dew. Then he said, Very sorry to do this Doctor, but I have to arrest you for the Murder & Mutilation of your wife.
Well, that word Mutilation (even nicely put by the Inspect) quite offended me, since it is a word foreign to a medical man. Then he handcuffed me.
I said, Cant she & me be together, the Insp. said he was very sorry, against the rules. Which means that the next time we will see each other will be in – what do they call it – the Dock. I dismiss it from my mind—
I said, For Miss Le Neve this is a very bad moment, will she be being treated with respect, please Inspector put my mind at rest.
Dew: Of course, I will have a word with the Captain to make sure she is looked after by 2 nurses from the ships crew, specially chosen with the right natures.
Me: I know Im the one that got her into that get-up, but can something be done to get her to look right?
Dew: Dont you worry Doctor, I'll have a word with the Captain here & he'll get some 1st Class lady passengers together & make a collection of female apparel.
Thats a weight off my mind. Imagine, a petticoat from here, a pair of stays from there – its a strange world & no mistake. So long as they cant make her a figure of fun.
So here I am in one empty cabin, & her in another, while they search *our* cabin. To keep me company, I have got a hefty keeper, in case I try to jump overboard or something.
(A minute ago when I took this Diary from my pocket & said I wanted to write in it, the keeper looked at me very

puzzled, I guess it was a funny thing to want at a moment like this, but he didnt stop me.)

I asked him if he had known about the Insp. & he said Of course, the whole crew knew he was on his way across on the *Laurentic*, & (thanks to the Wireless Telegraphy between Capt. & Insp.) so does the whole world, every newspaper has been full of it for 10 days, didnt you see that crowd on the quay at Father Point, the neighbouring town of Rimousquee (cant spell it) is bursting with journalists etc, you are more famous than King George V or General Tom Thumb.

When the Insp. comes in, he will take this Diary from me, but I would like to finish it off. (I have always been tidy.)

Now, recalling those words in the *Daily Mirror* – ghastly, horrible, grisly, fiendish – words which are in a million mouths at this minute – I have an odd thought on this, which – BEING INNOCENT – I can write down as the Man in the Street, also as realistic medical man, with a Degree.

IN THE SAME WAY THAT THE UNKNOWN PERSON WHO DID THIS WILL BE FAMOUS WHEN THEY CATCH HIM – IN THAT SAME WAY *I* AM (AT THIS MINUTE) FAMOUS ALL OVER THE WORLD, BECAUSE THEY THINK *I* DID IT.

And what they mean by *it*, is not the fact that in my house some woman met her death at the hands of a man (these things happen all over the world, dozens every day, & they hardly get into the papers at all) – by *it*, they mean that the man (whoever he is) tried to dispose of the body *in this way*.

And thats where the cheap sensationalism comes in, all over the world – sheer vulgar horror at the gory details which any Surgeon goes through & thinks little of, working on a body that can feel *nothing*. I speak with commonsense.

Prejudice, thats what rules the world. Now Im a Doctor, & Doctors face facts, & as a Doctor I have read of what they call Bestiality, viz: a man having sexual intercourse with an animal. And I have a question to ask, & I dont think the Archbishop of Canterbury himself can come up with an answer.

Why can a man have sexual intercourse with any number of women & go Scot-free, but if he *kills* one, he becomes a

criminal? Then – looking at it the opposite way – why, if the same man has sexual intercourse with a sheep, will he go to jail for years – but yet he is encouraged to kill that same sheep, carve it up, & *eat it*? Ever heard of mutton, Archbishop?
But thats just a philosophical thought. What is important is – HOW CAN I PROVE IM INNOCENT?
And yet you have to come back to one thing – thinking of my meeting with the Insp. earlier on – when he said *Good Morning Doctor* – it sounds strange this, but its true – behind the official nature of the thing, it was almost like two old friends meeting unexpectedly – hard to believe what the whole thing was about.
And that look I'd seen on him before, only this time more marked. Respect.
And I feel the same respect for him. Tracking me down like this, you have to hand it to them.
Come to think of it, when they took me here from the Capt.'s cabin, there was quite a crush of passengers outside – these things spread like wildfire – & they give me the same look. Respect, with a bit of fear thrown in, like for a tiger out in the Zoo on his way from one cage to another. Or the Royal Family.
People used to say, There goes little Crippen, he'll never got anywhere, poor chap. And now that THEY THINK I DID THIS TERRIBLE THING OF WHICH I AM INCAPABLE, Im Royalty. Thats funny.
Jump overboard? Funny, when I think of that nightmare I had last night (or whatever it was) when I was going to finish it off – now that it all has happened, I dont feel like that at all.
Because I am going to fight.
I'LL HAVE A TERRIBLE TIME PROVING MY INNOCENCE, but I am determined to do it, because my D must be protected *at all costs*.
And we'll be together once its over – we will *we will*. I'll never forget that last look she gave me, the love in it, & fear & shame. But most of all, the love.
Until then, I'll be in a series of cells just like this – when

not in that dock, fighting for my life, & hers.

So until she & me will be together again, theres nothing more to say.

Signed at Father Point, Sun. July 31, 1910.
H. H. Crippen.

Afterword

1

THE REST OF THE FACTS

ON THE AFTERNOON of Saturday 9 July, Inspector Dew had called at Albion House to find that Crippen had not returned from lunch: he and Ethel Le Neve were already on their way to Dieppe. Dew then circulated, to all police stations, a detailed description of the missing Mrs Crippen. (Later criticized for not issuing a description of Crippen instead, he was to protest that in this – representing Scotland Yard – he had been correct, for concerning Crippen there could only be surmise.)

Dew and Mitchell spent Monday 11 July – two days later – going through 39 Hilldrop Crescent with a fine tooth comb. It turned out to be an exhausting and fruitless day's work: they found nothing.

But – almost as if drawn by a subconscious afterthought – two days later Dew returned with Mitchell and they concentrated on the cellar. The 13th (July, this time) would seem to be, for Crippen, an unlucky date.

Studying the bricks and paving left clear of the old chandelier and the coal-dust and rubbish, Dew noticed – for the first time – that one of the bricks looked almost loose. He straightened up and fetched from the adjacent kitchen a small poker, and was able to prise up the brick. Underneath he found a layer of clay and lime; on disturbing this, he was overpowered by a stench which drove him and Mitchell up into the summer air for a moment's relief.

There can have been few moments in any man's life when

acute physical revulsion can have been so mingled with the intense excitement of success.

Three days later, on 16 July, Scotland Yard issued a warrant, WANTED FOR THE MURDER OF ... The Inspector had lost no time: the 'Police Bill' which was circularized immediately included a detailed description of the couple, their possible wardrobe and a full account of the jewellery.

There was, however, no mention of the strong possibility that the wanted lady was dressed as a boy; fortunately for the fugitives, William Long the 'dental mechanic' (possibly from fear of being suspected as an accomplice) had not yet vouchsafed to Dew the story of his shopping trip.

All ports were to be watched: but again fortunately for Mr John and Master Robinson, Antwerp was clearly not as vigilant as it might have been. Four days later, on 20 July, they boarded S.S. *Montrose* with no trouble.

But their luck ran out. Captain Kendall would appear to have been not only an amateur sleuth, but (judging by his long message by wireless telegraphy, to Scotland Yard) a sharp-eyed and resourceful one.

After studying the Police Bill, which had arrived on board with his papers, on the stroll mentioned by Crippen he *did* register that Master Robinson had held his father's hand and 'squeezed it immoderately'. (One doubts if, since that historic radio message, the word *immoderately* has ever been spelt out across the Atlantic.)

He also noticed that the lad's trousers were 'very tight about the hips, and split a bit down the back and secured with very large safety pins.' He then proceeded to challenge comparison with Sherlock himself: recalling that the official description of Crippen had included mention of glasses and false teeth, he observed (while entertaining the couple) that 'the mark on the nose, caused through wearing spectacles, has not worn off since coming on board'; next he proceeded to tell Crippen 'a story to make him laugh heartily, to see if he would open his mouth wide enough for me to ascertain if he had false teeth. This ruse was successful'. Did it cross the Captain's mind that 'this ruse' was being played on a professional dentist?

THE REST OF THE FACTS 169

THE PRISONERS WERE taken from S.S. *Montrose* to Quebec, and then extradited to England on S.S. *Megantic*, arriving at Liverpool on 28 August.

The Crippen trial, at the Central Criminal Court, Old Bailey (London), opened on 18 October (1910) and lasted five days, closing on 22 October.

The accused, in the teeth of glaring evidence, pleaded 'Not Guilty' and maintained, to his last violent gasp, the demeanour of courteous and kindly forbearance which had characterized him through life: except during the week of Monday 31 January 1910, when – at Number 39 – he had found himself cornered into a programme of industrious physical activity which was to miss complete success only by a hair's breadth – or rather by the breadth of a thin poker fetched from a kitchen.

At Pentonville Prison, London, at 9 am on 23 November (1910) Hawley Harvey Crippen was hanged. From that day to this, his name has flourished world-wide with a steadily enduring life: a legend evoking an incongruous double image. The Little Man and the Monster.

FOUR DAYS AFTER Crippen was sentenced to death, Ethel Le Neve was taken from Holloway Prison, tried (in the same court) for having been an accessory after the fact, defended by F. E. Smith (later Lord Birkenhead) and acquitted. She then disappeared into an obscurity which – by 1954, forty-four years later – must have seemed complete and permanent.

In that year, during which the *Sunday Dispatch* (London) was serializing Ursula Bloom's 'factual novel' *The Girl Who Loved Crippen*, its editor Charles Eade was visited by an irate Mr Neave: 'I am Ethel Le Neve's brother, and how dare you ....'

Miss Bloom went out of her way to meet him, and tactfully suggested that he might help her with the rest of her book. A week or two later she received a letter, signed 'Ethel Le Neve', suggesting that the brother was unsuitable as a consultant, and that Miss Bloom would do well to write to Miss Le Neve 'care of Mrs Smith' to an address in Addiscombe (a South London suburb) from which Mrs Smith would presumably forward the letter.

Miss Bloom did not write, having decided on a bolder move: she took a train to Addiscombe and walked to Mrs Smith's house, reaching it at between five and six on a fine summer afternoon: a semi-detached little house, in the usual endless row of identical dwellings. The door was opened by Mrs Smith.

Although the photographs extant of Ethel Le Neve had been taken a good forty-six years before, when Crippen's secretary was twenty-five, the visitor recognized the seventy-one-year-old lady facing her. Mrs Smith was Ethel Le Neve. And fortunate enough to have acquired the most invisible surname of all.

Miss Bloom said quickly, 'Please be reassured, I promise I am not here to invade your privacy.' Mrs Smith asked her in, and in the typical little front room, they sat facing each other.

With what must have been admirable sensitivity, the writer ventured into questionable territory with the caution of a doctor dealing with a special patient; starting as the tolerant visitor, she was to end up as the trusted friend. (Throughout the relationship, she never set eyes on Mr Smith; she gathered that he worked at Hampton's, the big West End furniture store, travelling up from Addiscombe each day, and that he physically resembled Crippen. His wife had borne him two children, a boy and a girl.)

Mrs Smith – of her own accord – talked freely of the past. (For example, the account at the beginning of Chapter 4, 'Love', of how Crippen met her in the street with her brothers and sisters, is her own, as told to Miss Bloom.) The day after Crippen's execution, 24 November 1910, she had left the country, under the name of Ethel Allen, on the S.S. *Majestic*, he having previously arranged for this to be financially possible; at his express wish, her destination was Canada. For her, this Atlantic crossing – her second within four months – must have been unbearably poignant.

At some time during the First World War, she returned to

London to be with a sister who was dying of consumption, and stayed with her old landlady Mrs Emily Jackson, who was still devoted to her. Then came marriage. And retirement to Addiscombe.

Her only social activity of any kind seems to have been to befriend a neighbour, a delicate little boy named Rex Dunning whom she nursed through an illness; he was to become as close to her as her own children, and when he had grown up she told him her secret.

By the end of 1966, the old lady was very frail, and on the point of admittance to hospital; one day she asked Rex Dunning to take her, that afternoon, for a London car-drive to two places she had known.

It was with some trepidation that he consented. Outside each of the two buildings, she sat in the car for several minutes, very upright, without speaking, then asked to be driven back home. The buildings were both prisons: Holloway and Pentonville. There had been no mention of Hilldrop Crescent.

Shortly afterwards, near the beginning of 1967, Mrs Ethel Smith died in Dulwich Hospital, London, S.E. 22, at the age of eighty-four. Thanks to Miss Bloom's persistently loyal refusal (sometimes under pressure) to disclose her friend's married name or whereabouts, the death passed unnoticed.

2

FACT AND FICTION IN THE DIARY

ALL THE RELEVANT factual aspects of the case, as set out in the indispensable *Trial of Hawley Harvey Crippen* in the 'Notable British Trials Series' (Williams Hodge and Company, 1920) are used in the composition of the Diary: these include dates, addresses, press headlines and reports, letters, telegrams and conversations mentioned at the trial. On these are based – to choose at random – such details as Crippen's being called 'Peter', and his informing Mrs Smythson that his dead wife was 'right up in the wilds of the mountains of California'. Moreover, every individual mentioned by name is a real person. (Including Mrs Crippen's 'admirer', Bruce Miller; by 1910 a husband and father in Chicago, he must have felt accutely embarrassed, six years after he had last seen her, to be summoned to London in order to appear in the witness-box at the most famous murder trial of the era.)

The only exception is Dr Masters (and family), Crippen's neighbour in Albion House; the reason for his fictional inclusion is obvious, his death serving to provide Crippen with the opportunity to hide the main (and incriminating) part of the Diary until its planned discovery in 1985.

THE FACTUAL BASIS of the Diary having been established, it is hardly necessary to admit that the diarist's thoughts, emotions, reasonings and judgements – on anything and everything – are surmised by the author. Also, any conversations which were not reported at the trial are equally reconstructions, by the author. These include all the exchanges between Ethel Le Neve and Crippen, also many between him and Inspector Dew, such as Crippen's mistaking the latter's phrase 'an' the seller' for 'in the cellar'.

IN MANY MURDER cases, there crops up at least one anomaly or discrepancy which becomes an exasperating puzzle. The Crippen dossier offers two, which have mystified students of the case for three-quarters of a century. I have thought it a good idea to take the opportunity – via the Diary – to suggest solutions for these two problems.

The first puzzle stems from Crippen's behaviour leading up to the night of 31 January 1910.

There is no doubt whatever, of course, that on or about that date, after his wife's violent death by poison Crippen buried her remains under the cellar of their house. But is it conceivable that a medical man of his commonsense and cunning (at a time when he presumably had his wits about him) would prepare a murder for that night, beforehand, by openly buying a deadly amount of poison from a well-known London pharmacy, and on that night deliberately administer a fatal dose to his victim, with the certain knowledge that he would be faced with the daunting task of disposing of the dead body, *in his own house*?

Surely what is more, is that while he could indeed have been planning to murder his wife, he had hatched the idea of doing so in a way which, with luck, would automatically rid him of the responsibility of a corpse: a way which would have been made feasible by a husband-and-wife voyage to Dieppe. Suddenly, on the fatal night of 31 January, through the accident of an overdose, he is faced with that same corpse. Instead of going to the police and making a clean breast of what has happened, he is overwhelmed by panic. With fatal results.

It is hard to believe that as he gradually sank into the quicksands of conclusive evidence, he did not – even at the eleventh hour – recant and confess the ironic manner of his wife's death; even with the jury's (and the public's) ineradicable horror of physical 'mutilation and dismemberment', his gesture might have saved him from the gallows. The reason he could not choose to make that gesture is even more incredible: the choice he had already made, and to which he obstinately clung to the last, was to maintain – against conclusive evidence – that the remains in the cellar were *not those of his wife*.

The second enigma is even harder to solve: the question of the pyjama-top found with the remains. A discovery which, as much as anything else, was to get Crippen hanged.

How could he, after his horrifying but understandable obsession with making what would be left of the body as unidentifiable as possible, for his own protection – how could he, at the last minute, have tossed into the grave a patently traceable scrap of his own clothing? The theory put forward in *Doctor Crippen's Diary* – that his action was a gesture, a savage and childishly ungovernable caprice – is based on a study of the psychology of murderers between the murder (or murders) and ultimate capture.

The planning during the first crucial hours (or even days and weeks) is frequently cool and brilliantly meticulous; but with success, the realism becomes, sporadically, less absolute.

In other words, the crime would seem to go to the criminal's head. The inventive and crafty devotion to detail, from day to day, goes side by side with a growing disregard for danger – a disregard stemming from self-satisfaction ripening into megalomania ('I can do no wrong'). The result is, more often than not, fatal.

This was particularly marked in the Moors Murders (1963-1965); as one crime after another remained unsuspected, the principal criminal showed a growing contempt for prudence, leading to a deliberate manufacture, and preservation, of evidence which was to become, in the end, damning.

And so it was with Crippen. Foolishness led to foolishness: such as his senseless insistence on Ethel Le Neve wearing his dead wife's jewellery and furs in public: his assumption that his wife's friends would accept the fact that a letter from her to them had been dictated to him and not even signed by her: his dismissal of their mounting suspicions: his inner conviction (until it was too late) that he was hoodwinking Scotland Yard: his asking a friend to shop for a boy's wardrobe (surely the most bizarre and foolhardy errand ever thought up by a guilty man trying to cover up his tracks) when he could so easily have waited and done his own shopping in Dieppe: his childlike trust in Captain Kendall's friendly overtures: all following (and fatally consonant with) the burying of the pyjama-top.

It is the phenomenon of a man who one moment is looking at life through a telescope, with a beautiful and abnormal clarity – and the next moment has his arm jolted and sees the whole prospect turn into an instant fuzzy blur.

In the end, to him the delusion seems reality, and the facts a mirage. For instance, take his midnight panic on the boat, when his dreadful position is suddenly starkly clear to him: next morning, once he has recovered from this and is back in his false world, his frantic facing of facts the night before turns, for him, into a regrettable lapse from common sense, and the renewed delusion becomes a sensible appraisal of a difficult situation. His world is – at last – inside-out.

3

THE WOMAN IN THE CASE

Since we now have Miss Bloom's previously undivulged disclosures this is the moment to examine an all-important issue: the role, in the Crippen story, of Ethel Le Neve.

Was she, as was mostly taken for granted by an impulsively censorious world, an unscrupulous accomplice to a brutal murder? Or . . . was she innocent?

The conclusion one has to come to, is a startling one; not only was she innocent, she *believed Crippen to be innocent*. This is borne out (a) by a letter in his hand and (b) by her long and reasoned conversations with Miss Bloom.

The letter, the last ever written by the condemned man, is the one he addressed to Ethel Le Neve on the eve of his execution, following the news that there was no hope of a reprieve. It is remarkable for more than one reason.

Firstly, for its sudden and unconvincing appeal to the Deity; secondly, for its moving bravery in the face of inevitable official annihilation (it is noteworthy that Major O. E. M. Davies, the then Governor of Pentonville Prison where Crippen was held, is the only such official in British prison history to have called personally on the Home Secretary to request a reprieve); thirdly, for its proof of unimpaired and selfless devotion to the love of his life; lastly (and most importantly), for his assumption of his own innocence. It is an assumption unequivocal and unshakeable.

> . . . God indeed must hear our cry to Him for Divine help in this last farewell . . . The Governor brought me the dreadful news at about 10 o'clock. He was most kind, & left me with, *God bless you! Good night* . . . When he had gone I first kissed your face in the photo, my faithful

devoted companion in all this sorrow . . . How am I to endure to take my last look at your dear face? . . . I know you will be the only one to mourn for me – but do not, dearest, think I expect you to put on mourning – that I leave you to decide on . . . I feel sure God will let my spirit be with you always: . . . & after this earthly separation will join our souls for ever . . .

The rest of the letter (that is, the bulk of it) consists of an intricate and fairly unintelligible examination of the medical evidence at the trial regarding the remains found under the cellar.

Now it is plain to everyone that . . . the fact that no navel was found on that piece of skin . . . is proof beyond any possible doubt that the remains found at Hilldrop Crescent were not those of B.E. . . . I write these things in the hope that the unreliability of the case brought against me, may be understood by thoughtful people . . .

It is true that Crippen knew the letter would be scrutinized by the authorities: but it is extremely unlikely that he hoped that such a protest might raise doubts as to his guilt, and at the eleventh hour persuade the Home Secretary to change his mind. The alternative theory is even more untenable: that Ethel Le Neve had known the facts, and that this was an attempt on his part to convince the authorities that she hadn't.

No, she believed him innocent; and he was not, at the last moment, going to make her crippled future even more of a ruin – in her own eyes – by the shock of a confession.

How COULD SHE believe him innocent, after sitting in that court, hour after hour, and hearing the evidence? Could she ever have imagined that, during the few hours a week when Number 39 was empty, a stranger had smuggled into the house the remains of an unknown human being and buried them under the cellar, *plus Crippen's pyjama-top?*

The answer is that, when subject to one of the two emotional forces which rule our world, sex and religion, a human being will believe anything. From the moment when Scotland Yard coaxed that first brick out of a disturbed floor, life – for Crippen's mistress – was to become a personal hell, and her only chance of survival was to have faith. To believe the unbelievable.

In this, of course, she was enormously helped by Crippen's complete credibility: indeed, it would not be surprising if, by the time he wrote that last letter to her, he himself had become so possessed by his delusions that he believed that the night of 31 January had not happened.

Ethel Le Neve's certainty of his innocence is completely borne out by Miss Bloom. In their talks, the former never wavered, for a second, in the face of proved facts; she had an answer to everything.

Q: How do you explain the fact that Crippen's pyjama-top was found with the remains?

A: It was not his: pyjamas identical with his had been sold in their thousands, and in that district. (She was presumably not reminded that although Inspector Dew found the trousers of Crippen's pyjama-suit neatly folded in a drawer upstairs, *the corresponding jacket was not in the drawer*. Would her rejoinder have been: 'The police must have destroyed it'?)

Q: If Crippen's wife was indeed alive at the time of his arrest – when the whole civilized world was buzzing with the case, and it was inevitable she should know of it – why did she not come forward to save an innocent man from the gallows?

A: His wife stayed silent *on purpose*. (Ethel Le Neve assured Miss Bloom that Crippen had told her he had heard his wife speak the words 'I'd like to see him hanged'.)

At that time, apart from treason, an individual could be hanged only for murder. Assuming that Mrs Crippen spoke these words, what murder could she possibly be envisaging – encouraged, presumably, by a top-ranking fortune-teller – except her own? The idea is unbelievable.

What *is* believable is that Mrs Crippen, during their

quarrels, had expressed a hope that her husband would 'come to a bad end', and that Crippen, probably more than once, had told Miss Le Neve of this. For her, it would later be only one step further, on the road of wishful thinking, to persuade herself that he had used the word 'hanging'. Anything to shield him.

Q: If the remains were not those of Crippen's wife, whose were they?

A: They were the remains of a person who had been murdered before the Crippens moved in in 1905. Crippen had told Miss Le Neve that when they did move in, they immediately noticed something which neighbours had commented on before them: hovering over the basement of Number 39, there was a constant and offensive smell.

One can hear Crippen's persuasive voice recounting all this to the bewildered girl. (I have placed this conversation in the Hotel des Ardennes, Brussels, when he felt he had to break to her the news of the discovery of 'a body'.)

She must have wondered, if only for a moment, (a) why nothing had ever been done to locate the source of an intolerable nuisance; (b) why the Crippens had chosen, as their dining-room, not a room further away from the basement, nearer the top of the house, but the one next to the offending cellar; and (c) why, five years later, when Scotland Yard was scouring the house for evidence of possible foul play, and noses as well as eyes were on the alert, the smell had – instead of getting more and more offensive – mysteriously evaporated. No, these would have been unbearably nagging doubts, and she turned her back on them.

POOR ETHEL LE NEVE. And poor Crippen. She loved him, he loved her. As a step towards diminishing the legend of the Monstrous Little Man, it would be fair to remember that.